Virtual Topology and Functor Geometry

PURE AND APPLIED MATHEMATICS

A Program of Monographs, Textbooks, and Lecture Notes

LECTURE NOTES IN PURE AND APPLIED MATHEMATICS

Recent Titles

G. Da Prato and L. Tubaro, Stochastic Partial Differential Equations and Applications

W. A. Carnielli et al., Paraconsistency

A. Benkirane and A. Touzani, Partial Differential Equations

A. Illanes et al., Continuum Theory

M. Fontana et al., Commutative Ring Theory and Applications

D. Mond and M. J. Saia, Real and Complex Singularities

V. Ancona and J. Vaillant, Hyperbolic Differential Operators and Related Problems

G. R. Goldstein et al., Evolution Equations

A. Giambruno et al., Polynomial Identities and Combinatorial Methods

A. Facchini et al., Rings, Modules, Algebras, and Abelian Groups

J. Bergen et al., Hopf Algebras

A. C. Krinik and R. J. Swift, Stochastic Processes and Functional Analysis: A Volume of Recent Advances in Honor of M. M. Rao

S. Caenepeel and F. van Oystaeyen, Hopf Algebras in Noncommutative Geometry and Physics

J. Cagnol and J.-P. Zolésio, Control and Boundary Analysis

S. T. Chapman, Arithmetical Properties of Commutative Rings and Monoids

O. Imanuvilov, et al., Control Theory of Partial Differential Equations

Corrado De Concini, et al., Noncommutative Algebra and Geometry

A. Corso, et al., Commutative Algebra: Geometric, Homological, Combinatorial and Computational Aspects

G. Da Prato and L. Tubaro, Stochastic Partial Differential Equations and Applications – VII

L. Sabinin, et al., Non-Associative Algebra and Its Application

K. M. Furati, et al., Mathematical Models and Methods for Real World Systems

A. Giambruno, et al., Groups, Rings and Group Rings

P. Goeters and O. Jenda, Abelian Groups, Rings, Modules, and Homological Algebra

J. Cannon and B. Shivamoggi, Mathematical and Physical Theory of Turbulence

A. Favini and A. Lorenzi, Differential Equations: Inverse and Direct Problems

R. Glowinski and J.-P. Zolesio, Free and Moving Boundries: Analysis, Simulation and Control

S. Oda and K. Yoshida, Simple Extensions with the Minimum Degree Relations of Integral Domains

I. Peeva, Syzygies and Hilbert Functions

V. Burd, Method of Averaging for Differential Equations on an Infinite Interval: Theory and Applications

F. Van Oystaeyen, Virtual Topology and Functor Geometry

Virtual Topology and Functor Geometry

Fred Van Oystaeyen
University of Antwerp
Belgium

Chapman & Hall/CRC
Taylor & Francis Group
Boca Raton London New York

Chapman & Hall/CRC is an imprint of the
Taylor & Francis Group, an informa business

Chapman & Hall/CRC
Taylor & Francis Group
6000 Broken Sound Parkway NW, Suite 300
Boca Raton, FL 33487-2742

10 9 8 7 6 5 4 3 2 1

International Standard Book Number-13: 978-1-4200-6056-0 (Softcover)

Library of Congress Cataloging-in-Publication Data

Oystaeyen, F. Van, 1947-
Virtual topology and functor geometry / Fred Van Oystaeyen.
p. cm. -- (Lecture notes in pure and applied mathematics ; 256)
Includes bibliographical references and index.
ISBN 978-1-4200-6056-0
1. Categories (Mathematics) 2. Grothendieck categories. 3. Representations of congruence lattices. 4. Sheaf theory. 5. Dynamics. 6. Noncommutative function spaces. I. Title. II. Series.

QA169.O97 2007
512'.62--dc22 2007022732

Visit the Taylor & Francis Web site at
http://www.taylorandfrancis.com

and the CRC Press Web site at
http://www.crcpress.com

Contents

Foreword

In order to arrive at a version of Serre's global sections theorem in the noncommutative geometry of associative algebras, one is forced to introduce a noncommutative topology of Zariski type. Sheaves over such a noncommutative topology do not constitute a topos, but that is exactly the reason why sheaf theory in this generality can carry the essential noncommutative information generalizing to a satisfactory extent classical scheme theory. The noncommutativity forces, at places, a departure from set theory-based techniques resulting in a higher level of abstraction, because opens are not sets of points. Based on some intuition stemming mainly from noncommutative algebra and classical geometry, I strived for an axiomatic introduction of noncommutative topology allowing at least a minimalistic version of geometry involving actual "spaces" and not merely a mask for noncommutative algebra! Completely new problems appear already at the fundamental level, requiring new ideas that sometimes almost alienate a pure algebraist. Not all such ideas are completely developed here, often I restricted myself to bare necessities but left room for many projects ranging from the exercise level to possible research. The spirit of these notes is somewhat experimental reflecting the initial stage of the theory. This may occasionally result in a certain imbalance between novelty sections on new aspects of virtual topology and functor geometry on one hand versus well-established parts of noncommutative algebra on the other. In either case I tried to supply sufficient background material concerning localization theory or some facts on the classical lattice $L(H)$ of quantum mechanics for some Hilbert space H.

On the other hand, I included a few topics that are, at this moment, only important for some of the research projects. In recent years "research training" for so-called young researchers became a trendy topic, and several of the included projects might be viewed in such a framework; however, some projects mentioned are probably hard and essential for better development of the theory and its applications. Intrinsic problems related to sheafification over a noncommutative space are the main topic in Section 4.2 and represent the introduction of a dynamic version of noncommutative topology and geometry. Since this construction is strictly related to the "absence" of points or of "enough points" in the noncommutative spaces, the dynamic theory as defined here is an exclusively noncommutative phenomenon; it is trivialized in the commutative case where space, and its topology, is described by sets of points. While reading Section 4.3 the reader should maintain a physics point of view because a noncommutative model for "reality" is hinted at; I included some observations related to this "spaced time," resulting from recent interactions with several physicists, just as food for thought. I welcome all reactions and suggestions, for example, concerning the projects or the general philosophy of the topic.

F. Van Oystaeyen

Acknowledgments

Research in this work was financially supported by :

- An E.C. Marie Curie Network (RTN - 505078) LIEGRITS
- A European Science Foundation Scientific Programme: NOG

The author fittingly supported these projects in return.

I thank my students and some colleagues for keeping the fire burning somewhere, and the Department of Mathematics at the University of Antwerp for staying small, even after the fusion of the three former Antwerp universities.

I especially appreciated moral support from E. Binz, B. Hiley, and C. Isham; they shared their vast knowledge in both physics and mathematics with me, and by showing their interest, motivated me to further the formal construction of noncommutative topology.

Finally, thanks to my family for the life power line.

Introduction

Noncommutativity of certain operations in nature as well as in mathematics has been observed since the early development of physics and mathematics. For example, compositions of rotations in space or multiplication of matrices are well-known examples often highlighted in elementary algebra courses. More recently even geometry became noncommutative, and nowadays motivation for the consideration of intrinsically noncommutative spaces stems from several branches of modern physics, for example, quantum gravity, some aspects of string theory, statistical physics, and so forth. From this point of view it seems to be necessary to have a concept of space and its geometry that is fundamentally noncommutative even to the extent that one would not expect that its mathematics is built on set theory or the theory of topoi. On the other hand, some branches of noncommutative geometry realize the *noncommutative space* solely via the consideration of noncommutative algebras as algebras of functions on an undefined fantasy object called the *noncommutative variety or manifold.* Nevertheless this technique is relatively successful, and it allows a perhaps surprising level of geometric intuition combined with algebraic formalism either in the algebraic or differential geometry setup.

Further generalization may be obtained by conveniently replacing sheaf theory on the Zariski or real topologies by more abstract theoretical versions of it. In such a theory, the objects of interest on the algebraic level are either some types of quantized algebras or suitable C^*-algebras. The fact that noncommutativity may force a departure from set theoretic foundations creates a parallel development on the side of logic involving non-Boolean aspects as in quantum logic or *quantales* replacing Grothendieck's *locales*. The different points of view fitting the abstract picture sketched above do not seem to fit together seamlessly; in particular, some desired applications in physics do not seem to follow from spaceless geometry, even if some applications do exist already. For example, the symbiosis between quantales and C^*-algebras defies more general applicability for algebras of completely different type. We may now rephrase the ever-tantalizing dilemma as points or no points, that's the question!

On one hand, the introduction of a pointless geometry defined by posets with suitable operations extending the idea of a lattice to the noncommutative situation, with the partial order relation not necessarily related to set theoretic inclusion, seems to be very appropriate. After all, an abstract poset approach to quantum gravity seems to be at hand!

On the other hand, there are points in a pointless geometry! In fact, there are different kinds of points, and in specific situations a certain type of point is more available than another. The problem then arises whether a noncommutative topology, defined on a noncommutative space in terms of a noncommutative type of lattice that replaces the set of opens, is to some extent characterized by sets (!) of noncommutative points.

Observe that when defining the Zariski topology on the prime (or maximal) ideal spectrum of a commutative (Noetherian) ring, one actually defines the opens by specifying their points and the spectrum is defined before the topology. The difference between presheaf and sheaf theory is completely encoded in the relations between *sections on opens* and *stalks at points*. The sheafification functor may be the ultimate example of this interplay, its construction depends on consecutive limit constructions from basic opens to points by direct limits, and from points to arbitrary opens by inverse limits. Even in classical commutative geometry there is a difference when prime ideals of the ring are viewed as points of the spectrum or only maximal ideals are considered as such. However, at the basic level there is absolutely nothing to worry about because the type of rings considered, for example, commutative affine algebras over a field, are Jacobson rings (and Hilbert rings); that is, every prime ideal is determined by the maximal ideals containing it, and in fact it is the intersection of them. So even the commutative case learns that once a topology is given there are still several consistent ways to decide what the points, but when a notion of points is fixed first, the topology has to be adjusted to this notion in order to obtain a useful sheaf theory.

Another most important property in classical geometry is that varieties, schemes, or manifolds are locally affine in some sense; for example, every point has an affine neighborhood. In a pointless geometry the latter property is hard to understand and a serious modification seems to be necessary. It will turn out that for this reason, one has to introduce representational theoretic aspects in the abstract theory. Now, for the kind of noncommutative algebraic geometry in the sense of a generalization of scheme theory over noncommutative algebras, as promoted by the author (for example in [44]), the presence of module theory and a theory of quasicoherent sheaves make this possible.

But what remains if we try to drop all unnecessary (?) restrictions concerning the presence of an algebra, modules, spectra, points, and so forth, and try to arrive at a barely abstract geometry based on a kind of topology equipped with some functors on a general but suitable category or family of categories? Well, perhaps virtual topology and functor geometry! In the following I try to indicate how such a general theory will have to deal with the issues raised above.

First, noncommutative topology is introduced via the notion of a *noncommutative lattice* where the operations $\wedge$ and $\vee$ are defined axiomatically and they are less strictly connected to the partial order than the meet and join in usual lattices. The noncommutative topologies may be considered as sets of opens, but an open can in no way be viewed as a set. Noncommutative topologies do fit in a theory of noncommutative Grothendieck topologies but not in topos theory; a noncommutative version of the latter remains to be developed.

Then points and minimal points may be defined in a generalized Stone space associated to a noncommutative topology. There are not enough points to characterize an open to which they belong, but there is a well-behaved notion of commutative shadow of a noncommutative space, which is given by a real lattice in the usual sense, and where the commutative opens are characterized by sets of points. At this point generalized function theory could be developed but we did not go into this; rather we introduced sheaf theory on noncommutative topologies and verified that they transfer nicely to the generalized Stone space. A complete symmetrization of Grothedieck's definition of a Grothendieck topology leads to noncommutative (left, right, skew) versions of this, and the noncommutative topologies defined axiomatically fit into the latter framework by restricting to certain partial order relations, that is, the so-called generic relations. All of this is in Chapter 2 ending with two fundamental examples: the lattice of torsion theories or Serre quotient categories of a Grothendieck category, and the lattice of closed linear subspaces of a Hilbert space. The first one has enough points in the "prime" sense; the second has enough minimal points (maximal filters) in the Stone space.

In Chapter 3, Grothendieck categorical representations are studied with the aim of arriving at an abstract notion of *affine open*. When applying this to the algebraic geometry of associative algebras, for example, schematic algebras, their modules and the localizations of module categories, the general notion of affine open describes exactly the opens corresponding to exact localization functors commuting with direct sums (functors with an adjoint of a specific type). The general notion of quotient representation may then be used to explain how noncommutative projective spaces arise from noncommutaive affine spaces.

Some sheaf theory is developed; in a sense this is an extension of a theory of quantum sheaves considered earlier by changing from categories of opens in a topology to more general lattices, but now we even allow the suitable noncommutative version of lattices.

The creation of a new theory sometimes opens many doors, maybe too many doors. For example, the further development of the noncommutative version of topology, for example, closed sets, compactness, convergence structure is possible in the generalized Stone space, but we have not even tried to go there. Even though it may well be that such theory is interesting in its own right, we have only mentioned this as a project for the zealous reader looking for an original way to test his/her skills. Even more haunting ideas about noncommutative probability or measure theory have suffered the same fate. Some projects, however, are more straightforward exercises leading to possible research projects.

The final section starts out swinging — well at least we propose a dynamical version of topology and sheaf theory, providing at least one solution of the problem of sheafification independent of generalizations of topos theory. It required a rephrasing of continuities in a poset setup with a totally ordered set (time!) as a parameter set. The result is a spectrum with a classical topology existing at each moment but not varying in time the way the noncommutative topologies do. This may be seen as a mathematical uncertainty principle or better as mathematics of observation. One might hope that physical phenomena, in particular quantum theories, may suitably be phrased in terms of this observational mathematics — perhaps a dream.

For the more algebraic, or more geometric, applications of ideas expounded in these notes, we may refer to earlier work in noncommutative geometry — in particular the theory of schematic algebras (see [48]). It is not surprising that the geometric structures stripped to their naked abstraction retain a somewhat esoteric character, highlighting mainly partial ordered sets with noncommutative operations but related to lattices (via the commutative shadow of spaces), categorical methods and functorial constructions, a further abstraction of sheaf theory and spectral constructions, and a categorical representation theory using Grothendieck categories. Some ideas in these notes have already inspired a few recent papers in physics, so without trying to claim more, I hope that the exercise of digging deep for the abstract skeleton of geometry may lead to a further unification of different kinds of noncommutative geometry and point to an actual space of an intrinsically noncommutative nature, perhaps allowing the expression of observations concerning natural phenomena.

Projects

Chapter 1

A Taste of Category Theory

1.1 Basic Notions

We assume that the reader is familiar with the foundations of set theory, at least in its intuitive version.

Acategory $\underline{C}$ consists of a class of objects together with sets $\mathrm{Hom}_{\underline{C}}(X, Y)$, for any pair of objects X, Y of $\underline{C}$, satisfying suitable conditions listed hereafter. The elements of $\mathrm{Hom}_{\underline{C}}$ are called $\underline{C}$-morphisms or just morphisms, if there is no ambiguity concerning the category considered. For any object X of $\underline{C}$ there is a distinguished element $I_X \in \mathrm{Hom}_{\underline{C}}(X, X)$, called the **identity morphism of** X. For any triple X, Y, Z of objects in $\underline{C}$ there is a composition map:

$\mathrm{Hom}_{\underline{C}}(X, Y) \times \mathrm{Hom}_{\underline{C}}(Y, Z) \to \mathrm{Hom}_{C}(X, Z)$, $(f, g) \mapsto g \circ f$ such that the following properties hold:

i. For $f \in \mathrm{Hom}_{\underline{C}}(X, Y)$, $g \in \mathrm{Hom}_{\underline{C}}(Y, Z)$, $h \in \mathrm{Hom}_{\underline{C}}(Z, W)$ we have: $h \circ (g \circ f) = (h \circ g) \circ f$.

ii. For $f \in \mathrm{Hom}_{\underline{C}}(X, Y)$ we have $f \circ I_X = f = I_Y \circ f$.

iii. If $(X, Y) \neq (X', Y')$, then $\mathrm{Hom}_{\underline{C}}(X, Y)$ and $\mathrm{Hom}_{\underline{C}}(X', Y')$ are disjoint sets.

1.1.1 Examples and Notation

i. The category **Set** is obtained by taking the class of all sets using maps for the morphisms.

ii. The category **Top** is obtained by taking the class of all topological spaces using continuous functions for the morphisms.

iii. The category **Ab** is obtained by taking the class of all abelian groups using groups homomorphisms for the morphisms.

iv. The category **Gr** is obtained by taking the class of all groups using group homomorphisms for the morphisms.

v. The category **Ring** is obtained by taking the class of all rings using ring homomorphisms for the morphisms.

vi. For any given ring R the class of left R-modules using left R-linear maps for the morphisms defines the category R-**mod**. The category defined in a similar way but using right R-modules and right R-linear maps is denoted by **mod**-R.

Definition 1.1
Consider a class $\underline{\mathcal{D}}$ consisting of objects of $\underline{\mathcal{C}}$, then $\underline{\mathcal{D}}$ is said to be a **subcategory** of $\mathcal{C}$ when the following properties hold:

i. For objects X, Y in $\underline{\mathcal{D}}$ we have: $\mathrm{Hom}_{\underline{\mathcal{D}}}(X, Y) \subset \mathrm{Hom}_{\underline{\mathcal{C}}}(X, Y)$.

ii. Composition of morphisms in $\underline{\mathcal{D}}$ is as in $\underline{\mathcal{C}}$.

iii. For X in $\underline{\mathcal{D}}$, I_X is the same as in $\underline{\mathcal{C}}$.

We say that $\underline{\mathcal{D}}$ is a **full subcategory** of $\underline{\mathcal{C}}$ when $\mathrm{Hom}_{\mathcal{D}}(X, Y) = \mathrm{Hom}_{\mathcal{C}}(X, Y)$ for all X, Y of $\underline{\mathcal{D}}$. In the list of examples one easily checks that **Ab** is a subcategory of **Gr** but, for example, **Gr** is not a full subcategory of **Set**.

Definition 1.2
For a set S and a family of categories $(\underline{\mathcal{C}})_{s\in S}$ we define the **direct product** category $\underline{\mathcal{C}}$ by taking the class of objects to consist of the families $(X_s)_{s\in S}$ of objects X_s of $\underline{\mathcal{C}_s}$ for $s \in S$. If $X = (X_s)_{s\in S}$, $Y = (Y_s)_{s\in S}$ are such families, then $\mathrm{Hom}_{\underline{\mathcal{C}}}(X, Y) = \{(f_s)_{s\in S}, f_s \in \mathrm{Hom}_{\mathcal{C}_s}(X_s.Y_s), s \in S\}$. Composition of morphisms is defined componentwise. This direct product category $\underline{\mathcal{C}}$ will be denoted by $\prod_{s\in S}\underline{\mathcal{C}_s}$; in case $\underline{\mathcal{C}_s} = \underline{\mathcal{C}}$ for all $s \in S$, then we also write $\underline{\mathcal{C}}^S$. In case $S = \{1, \dots, n\}$ we also write $\underline{\mathcal{C}_1} \times \underline{\mathcal{C}_2} \times \cdots \times \mathcal{C}_n$ for the direct product.

This paragraph deals with specific properties of morphisms. A $\underline{\mathcal{C}}$-morphism from an object X to an object Y will be denoted by $X \to Y$, and if $f \in \mathrm{Hom}_{\underline{\mathcal{C}}}(X, Y)$ we will write $f : X \to Y$.

A **monomorphism** (in $\underline{\mathcal{C}}$) is a morphism $f : X \to Y$ such that for any object Z and given morphisms $h, g \in \mathrm{Hom}_{\underline{\mathcal{C}}}(Z, X)$ such that $f \circ h = f \circ g$ we must have $g = h$. Dually, an **epimorphism** is a morphism $f : X \to Y$ such that for any object W of $\underline{\mathcal{C}}$, and given morphisms $h, g \in \mathrm{Hom}_{\underline{\mathcal{C}}}(Y, W)$ such that $h \circ f = g \circ f$, we must have $h = g$. An **isomorphism** is a morphism $f : X \to Y$ for which there exists a morphism $g : Y \to X$ such that $g \circ f = I_X$, $f \circ g = I_Y$. In case a g as above exists, then it is unique, as one easily checks; it is called the inverse of f and will often be denoted by f^{-1}. In a straightforward way, one verifies that an isomorphism is necessarily an epimorphism as well as a monomorphism. Observe that a morphism that is both a monomorphism and an epimorphism need not necessarily be an isomorphism; indeed in **Ring** the canonical inclusion $\mathbb{Z} \hookrightarrow \mathbb{Q}$ is both a monomorphism and an epimorphism! Composition of monomorphisms, respectively epimorphisms, respectively isomorphisms, yields again a monomorphism, respectively epimorphism, respectively isomorphism. The duality between monomorphisms and epimorphisms may best be phrased by passing to the **dual category** $\underline{C}^o$, having the same class of objects as $\underline{C}$ but with $\text{Hom}_{\underline{C}^o}(X, Y) = \text{Hom}_{\underline{C}}(Y, X)$ by definition.

Now a morphism $f : X \to Y$ in $\underline{C}$ is a monomorphism, respectively an epimorphism, if and only if f is an epimorphism, respectively a monomorphism when seen as a morphism $Y \to X$ in the dual category $\underline{C}^o$.

Definition 1.3: Subobjects in $\underline{C}$
In many examples and applications the objects considered need not be sets, hence a correct definition of the term *subobjects* requires some care. Fix an object X of $\underline{C}$; for any W of $\underline{C}$ we have a set $\text{Mono}_{\underline{C}}(W, X)$ consisting of monomorphisms $W \to X$. For objects U and W of $\underline{C}$ we have a product set $\text{Mono}_{\underline{C}}(U, X) \times \text{Mono}_{\underline{C}}(W, X)$; if (α, β) is in the latter, then we may define $\alpha \leq \beta$ if there exists a morphism $\gamma : U \to W$ such that $\beta \circ \gamma = \alpha$. In case a γ as before exists, it is unique and also a monomorphism. We say that α and β are equivalent monomorphisms if $\alpha \leq \beta$ and $\beta \leq \alpha$; indeed, the foregoing defines an equivalence relation! Since we are dealing with sets now, we may evoke the Zermelo axiom and choose a representative in every equivalence class of monomorphisms. The resulting monomorphism is called a **subobject of** X in $\underline{C}$. **Quotient objects** may now also be defined in a formally dual way by passing from $\underline{C}$ to $\underline{C}^o$. It is not difficult to verify that a subobject of a subobject is again a subobject and a quotient object of a quotient object is a quotient object.

Definition 1.4: Initial and Final Object
An object I, respectively F, of $\underline{C}$ such that $\text{Hom}_{\underline{C}}(I, X)$, respectively $\text{HOM}_{\underline{C}}(X, F)$, is a singleton for every X of C is called an **initial object** of $\underline{C}$, respectively a **final object** of $\underline{C}$. Two initial, respectively final, objects of $\underline{C}$ are necessarily isomorphic. A **zero object** of $\underline{C}$ is an object that is initial and final. This allows us to distinguish zero morphisms as those $f : X \to Y$ that factorize through the (unique up to isomorphism) zero object. If a zero object exists, then we denote it by O; then each set $\text{Hom}_{\underline{C}}(X, Y)$ has precisely one zero morphism denoted o_{XY} or just o when no confusion can arise.

Definition 1.5: Product and Coproduct Objects
To a family $(X_s)_{s \in S}$ of objects in $\underline{C}$ we may associate the **product** $\prod_{s \in S} X_s = X$ if we can solve a universal construction problem in $\underline{C}$. The object X we look for should come equipped with a family of morphisms $(\pi_s)_{s \in S}, \pi_s : X \to X_s$ for $s \in S$, such that for any object Y of $\underline{C}$ with given morphisms $f_s : Y \to X_s$ for $s \in S$, there exists a unique morphism $f : Y \to X$ such that $\pi_s \circ f = f_s$ for all $s \in S$. If an object X with these properties does exist in $\underline{C}$ then it is unique up to isomorphism and we use the notation $X = \prod_{s \in S} X_s$. In case $S = \{1, \ldots, n\}$ then we also write $X = X_1 \times \cdots \times X_n$ (suppressing $\underline{C}$ in the product notation).

The notion of **coproduct** is defined dually. If the coproduct of a family $(X_s)_{s \in S}$ exists we will denote it by $\coprod_{s \in S} X_s$; in case $S = \{1, \ldots, n\}$ it is customary to write it as $X_1 \oplus \cdots \oplus X_n$.

Now that our categories need not be sets, we may not be able to talk about maps from one category to another, notwithstanding the fact that we may without any problem *associate* an object of a category to an object in another category. For categories $\underline{\mathcal{D}}$ and $\underline{\mathcal{C}}$ we let a **covariant functor** F from $\underline{\mathcal{D}}$ to $\underline{\mathcal{C}}$ be defined by associating it to an object X of $\underline{\mathcal{D}}$ an object $F(X)$ of $\underline{\mathcal{C}}$ and to a morphism $f : X \to Y$ in $\underline{\mathcal{D}}$ a morphism $F(f) : F(X) \to F(Y)$ in $\underline{\mathcal{C}}$ such that the following properties hold:

i. $F(I_X) = I_{F(X)}$ for every X of $\underline{\mathcal{D}}$.

ii. $F(g \circ f) = F(g) \circ F(f)$ for $f : X \to Y, g : Y \to Z$ in $\underline{\mathcal{D}}$.

A **contravariant functor** from $\underline{\mathcal{D}}$ to $\underline{\mathcal{C}}$ is then just a covariant functor $\mathcal{D}^o \to \underline{\mathcal{C}}$. Usually, when F is a covariant functor from $\mathcal{D}$ to $\underline{\mathcal{C}}$, set-theoretic-inspired notation is used to express this by $F : \underline{\mathcal{D}} \to \underline{\mathcal{C}}$; the context makes it clear that we do not mean to imply by this that F is a map!

Definition 1.6: Full and Faithful Functors
A covariant functor $F : \underline{\mathcal{D}} \to \underline{\mathcal{C}}$ yields for any X, Y in $\underline{\mathcal{C}}$ a map $\mathrm{Hom}_{\underline{\mathcal{D}}}(X, Y) \to \mathrm{Hom}_{\underline{\mathcal{C}}}(F(X), F(Y))$, which we denote also by: $f \mapsto F(f)$.

We say that F is **faithful**, respectively, **full**, respectively full and faithful, if the above map $f \mapsto F(f)$ is injective, respectively surjective, respectively bijective. Note that for any category $\underline{\mathcal{C}}$ there exists an identity functor $1_{\underline{\mathcal{C}}} : \underline{\mathcal{C}} \to \underline{\mathcal{C}}$ defined by $1_{\underline{\mathcal{C}}}(X) = X$ for every object X of $\underline{\mathcal{C}}$, and $1_{\underline{\mathcal{C}}}(f) = f$ for every morphism $f \in \mathrm{Hom}_{\underline{\mathcal{C}}}(X, Y)$. Obviously, the functor $1_{\underline{\mathcal{C}}}$ is always full and faithful.

Can functors between categories make up a category, and then what should be the morphisms? We do not treat functor categories in depth here but restrict ourselves to recalling a few fundamental notions related to this idea.

Definition 1.7: Functorial Morphisms
Consider a pair of covariant functors $F, G : \underline{\mathcal{D}} \to \underline{\mathcal{C}}$. A **functorial morphism** $\varphi : F \to G$ is given by morphisms $\varphi(X) : F(X) \to G(X)$ for X an object of $\underline{\mathcal{D}}$, such that for $f : X \to Y$ in $\underline{\mathcal{D}}$ we have: $\varphi(Y) \circ F(f) = G(f) \circ \varphi(X)$; in other words we have a commutative diagram of morphisms in $\underline{\mathcal{C}}$:

$$\begin{array}{ccc} F(X) & \xrightarrow{F(f)} & F(Y) \\ {\scriptstyle \varphi(X)}\downarrow & & \downarrow{\scriptstyle \varphi(Y)} \\ G(X) & \xrightarrow[G(f)]{} & G(Y) \end{array}$$

In case $\varphi(X)$ is an isomorphism for all objects X of $\underline{\mathcal{D}}$, φ is said to be a **functorial isomorphism** and we denote this by $F \simeq G$.

For functorial morphisms $\varphi : F \to G$ and $\psi : G \to H$ the composition $\psi \circ \varphi : F \to H$ may be defined by $(\psi \circ \varphi)(X) = \psi(X) \circ \varphi(X)$ for all X, and this yields again a functorial morphism. Let $\mathrm{Hom}(F, G)$ stand for the class of functorial morphisms from F to G. There exists an identity functorial morphism $1_F : F \to F$ defined by putting $1_F(X) = 1_{F(X)}$ for all X of $\underline{\mathcal{D}}$. Since $\mathrm{Hom}(F, G)$ need not be a set in general, we meet a small problem here in fitting this again in the framework of a category where the morphisms between two objects should be a set. In case $\underline{\mathcal{D}}$ is a small category, then $\mathrm{Hom}(F, G)$ is also a set.

Definition 1.8: Equivalences and Dualities
The covariant functor $F : \underline{\mathcal{D}} \to \underline{\mathcal{C}}$ is said to be an **equivalence** of categories when there exists a covariant functor $G : \underline{\mathcal{C}} \to \underline{\mathcal{D}}$ such that $G \circ F \simeq 1_{\underline{\mathcal{D}}}$ and $F \circ G \simeq 1_{\underline{\mathcal{C}}}$. In case $G \circ F = 1_{\underline{\mathcal{D}}}$ and $F \circ G = 1_{\underline{\mathcal{C}}}$, F is called an **isomorphism of categories**, and in that case $\underline{\mathcal{D}}$ and $\underline{\mathcal{C}}$ are said to be isomorphic categories. A contravariant functor $F : \underline{\mathcal{D}} \to \underline{\mathcal{C}}$ defining an equivalence between $\underline{\mathcal{D}}^o$ and $\underline{\mathcal{C}}$ is said to be a **duality of categories**, and in that situation $\underline{\mathcal{D}}$ and $\underline{\mathcal{C}}$ are said to be dual.

Theorem 1.1
A covariant functor $F : \underline{\mathcal{D}} \to \underline{\mathcal{C}}$ is an equivalence if and only if:

i. F is full and faithful.

ii. For any $Y \in \underline{\mathcal{C}}$ there is an $X \in \underline{\mathcal{D}}$ such that $Y \simeq F(X)$.

To an object X of $\underline{\mathcal{C}}$ we may associate a contravariant functor $h_X : \underline{\mathcal{C}} \to \underline{\mathrm{Set}}$ by putting $h_X(Y) = \mathrm{Hom}_{\underline{\mathcal{C}}}(Y, X)$ and for a morphism $f : Y \to Z$ in $\underline{\mathcal{C}}$ we define $h_X(f) : h_X(Z) \to h_X(Y)$ by $h_X(f)(z) = z \circ f$ for any $z \in h_X(Z)$.

*A functor $F : \underline{\mathcal{C}} \to \underline{\mathrm{Set}}$ is said to be **representable** if there is an object X of $\underline{\mathcal{C}}$ such that F is isomorphic to the functor $\mathrm{Hom}_{\underline{\mathcal{C}}}(X, -) = h_X$.*

Theorem 1.2: The Yoneda Lemma
For objects A and B of $\underline{\mathcal{C}}$ there exists a natural bijection of $\mathrm{Hom}(h_A, h_B)$ to $\mathrm{Hom}_{\underline{\mathcal{C}}}(B, A)$. In particular $\mathrm{Hom}(h_A, h_B)$ is a set.

Corollary 1.1
*The category $\underline{\mathcal{C}}^o$ is isomorphic to the category of **representable functors** $\underline{\mathcal{C}} \to \underline{\mathrm{Set}}$ with the functorial morphisms for the morphisms.*

1.2 Grothendieck Categories

The categories appearing in algebraic geometry, be it commutative or not, have very special properties, for example, modules over a ring, graded modules over a graded ring and so forth. For several results the class of abelian categories is suitable, but the best behaved categories we shall use are the so-called Grothendieck categories. These are rather close to being categories of left modules over a ring; the extra generality allows us to include categories of graded modules as well as certain categories of presheaves or sheaves.

A category $\underline{\mathcal{C}}$ is **pre-additive** if the following three properties hold:

i. For X, Y in $\underline{\mathcal{C}}$, $\mathrm{Hom}_{\underline{\mathcal{C}}}(X, Y)$ is an abelian group with zero element O_{XY} called the zero morphism.

ii. For X, Y, Z in $\underline{\mathcal{C}}$ and f, f_1, f_2 in $\mathrm{Hom}_{\underline{\mathcal{C}}}(X, Y)$, g, g_1, g_2 in $\mathrm{Hom}_{\underline{\mathcal{C}}}(Y, Z)$, we obtain:

$$g \circ (f_1 + f_2) = g \circ f_1 + g \circ f_2$$
$$(g_1 + g_2) \circ f = g_1 \circ f + g_2 \circ f$$

iii. There is an object X of $\underline{\mathcal{C}}$ such that $1_X = O_{XX}$. Clearly such X is a zero object, unique up to isomorphism, usually denoted by O.

It is obvious that the dual of a pre-additive category is again pre-additive. A functor between pre-additive categories may have some additivity properties, too; for example, we say that $F : \underline{\mathcal{D}} \to \underline{\mathcal{C}}$, where $\underline{\mathcal{D}}$ and $\underline{\mathcal{C}}$ are pre-additive, is an **additive functor** if for $f, g \in \mathrm{Hom}_{\underline{\mathcal{D}}}(X, Y)$, where X, Y are objects of $\underline{\mathcal{D}}$, we have: $F(f + g) = F(f) + F(g)$.

If $O_{\mathcal{D}}$ is the zero object if $\underline{\mathcal{D}}$, then $F(O_{\underline{\mathcal{D}}})$ is the zero object of $\underline{\mathcal{C}}$, say $O_{\underline{\mathcal{C}}}$.

An additive category is a pre-additive category $\underline{\mathcal{C}}$ such that for any two objects of $\underline{\mathcal{C}}$ a coproduct exists in $\underline{\mathcal{C}}$.

Definition 1.9: Abelian Categories

An additive category $\underline{\mathcal{C}}$ is said to be an **abelian category** if it satisfies conditions **AB.1** and **AB.2**:

AB.1

For any morphism $f : X \to Y$ in $\underline{\mathcal{C}}$, both $\mathrm{Ker}(f)$ and $\mathrm{Coker}(f)$ exist in $\underline{\mathcal{C}}$; then f may be decomposed as indicated in the following diagram:

$$\begin{array}{ccccccc} \mathrm{Ker}(f) & \xrightarrow{i} & X & \xrightarrow{f} & Y & \xrightarrow{\pi} & \mathrm{Coker}(f) \\ & & \downarrow{\scriptstyle\lambda} & & \downarrow{\scriptstyle\mu} & & \\ & & \mathrm{Coim}(f) & \xrightarrow[\overline{f}]{} & \mathrm{Im}(f) & & \end{array}$$

where $f = \mu \circ \overline{f} \circ \lambda$ and i and μ are monomorphisms and π, λ are epimorphisms.

AB.2

For every f as in **AB.1**, $\overline{f}$ is an isomorphism. In any category verifying **AB.2**, a morphism is an isomorphism exactly when it is both a monomorphism and an epimorphism.

Definition 1.10: Exact Sequences and Functors

Suppose that $\underline{\mathcal{C}}$ is pre-additive such that **AB.1.** and **AB.2.** hold. A sequence of morphism $X \xrightarrow[f]{} Y \xrightarrow[g]{} Z$ in $\underline{\mathcal{C}}$ is **exact** if $\mathrm{Im}(f) = \mathrm{Ker}(g)$ as subobjects of Y. An arbitrary (long) sequence is said to be **exact** if every subsequence of two consecutive morphisms is exact in the sense defined above. An additive functor $F : \underline{\mathcal{D}} \to \underline{\mathcal{C}}$, where both categories are pre-additive and such that **AB.1** and **AB.2** hold, is said to be **left exact**, and respectively **right exact** if for any exact sequence of morphisms in $\underline{D}$:

$$0 \to X \to Y \to Z \to 0$$

the following sequence is exact:

$$\begin{array}{lccccccc} & 0 & \to & F(X) & \to & F(Y) & \to & F(Z) & & \\ \text{respectively} & & & F(X) & \to & F(Y) & \to & F(Z) & \to & 0 \end{array}$$

When F is both left and right exact, then F is **exact**.

Consider X, Y in an additive category $\underline{C}$ and let $X \oplus Y$ be their coproduct. By definition of the coproduct there exist natural morphisms: $i_X : X \to X \oplus Y$, $\pi_X : X \oplus Y \to X$, $i_Y : Y \to X \oplus Y$, $\pi_Y : X \oplus Y \to Y$, such that $\pi_X \circ i_X = 1_X$, $\pi_Y \circ i_Y = 1_Y$, $\pi_X \circ i_Y = 0 = \pi_Y \circ i_X$, $1_{X \oplus Y} = i_X \circ \pi_X + i_Y \circ \pi_Y$. This actually establishes that $(X \oplus Y, \pi_X, \pi_Y)$ is a product of X and Y in $\underline{C}$. Consequently, if $\underline{C}$ is additive, respectively abelian, then $\underline{C}^o$ is too.

Lemma 1.1

A functor F between additive categories is an additive functor if and only if it commutes with finite coproducts.

*In [17] A. Grothendieck introduced several extra axioms on abelian categories, gradually strengthening the definition until the notion of the Grothendieck category, as we know it, appears. The axioms **AB.3**, **AB.4**, **AB.5** and their duals **(AB.3)***, **(AB.4)***, **(AB.5)*** are not independent; in fact, **AB.5** presupposes **AB.3** and implies **AB.4**. We just recall definitions and basic facts.*

AB.3

Arbitrary coproducts exist in $\underline{C}$.

(AB.3)*

Arbitrary products exist in $\underline{C}$. In case **AB.3** holds in $\underline{C}$ we may define for any nonempty set S a functor $\oplus_{s \in S} : \underline{C}^{(S)} \to \underline{C}$, associating to a family of $\underline{C}$-objects, indexed by S the coproduct (sometimes called the direct sum) of that family. The functor $\oplus_{s \in S}$ is always right exact.

AB.4

For any nonempty set S, $\oplus_{s \in S}$ is an exact functor.

(AB.4)*

For a nonempty set S, $\prod_{s \in S}$ is an exact functor.

If $\underline{C}$ is abelian and satisfies **AB.3.**, then for any family of subobjects $(X_s)_{s \in S}$ of X we may define a “smallest” subobject of X, denoted by $\sum_{s \in S} X_s$, such that all X_s are subobjects of the latter. The quotation marks around smallest refer to the fact that some care is necessary with the interpretation in view of the definition of subobject; compare Definition 1.3. The object $\sum_{s \in S} X_s$ is called the **sum** of $(X_s)_{s \in S}$. Dually, if $\underline{C}$ is an abelian category satisfying **(AB.3)***, then for every family $(X_s)_{s \in S}$ of subobjects of X we may associate $\cap_{s \in S} X_s$, the largest subobject of X contained in each X_s, $s \in S$. The subobject $\cap_{s \in S} S_s$ is called the **intersection** of $(X_s)_{s \in S}$. Observe that in any abelian category finite products do exist, hence the intersection of a finite family exists (it is enough to have an intersection of two objects).

AB.5
Let $\underline{C}$ be an abelian category satisfying **AB.3**. Consider an object X of $\underline{C}$ and subobjects X_s, $s \in S$, and Y, such that the family $(X_s)_{s\in S}$ is right filtered, then: $(\sum_{s\in S} X_s) \cap Y = \sum_{s\in S}(X_s \cap Y)$.

(AB.5)*
The dual of **AB.5**.

Observation 1.1
An abelian category such that **AB.3** and **AB.5** hold also satisfies **AB.4**.

Definition 1.11: Generators for an Abelian Category
Consider the family $(X_s)_{s\in S}$ in the abelian category $\underline{C}$; we say that $(X_s)_{s\in S}$ is a **family of generators** if for every object X and subobject $Y \neq X$ in $\underline{C}$ there is an $s \in S$ and a morphism $f : X_s \to X$ such that $\text{Im}(f)$ is not a subobject of Y. An object U of $\underline{C}$ is said to be a **generator** if $\{U\}$ is a family of generators.

Definition 1.12: Grothendieck Category
An additive category satisfying **AB, 1,** ... **AB.5**, and having a generator is a **Grothendieck category**. Observe that an abelian category, such that both **AB.5** and **(AB.5)*** hold, is necessarily the zero category (category consisting of the zero object with the zero morphism). Consequently the opposite of a Grothendieck category is never a Grothendieck category.

To end this section we recall some facts about adjoint functors. The notion of adjointness is very fundamental, and it has applications in different areas of mathematics.
Consider functors $F : \underline{C} \to \underline{D}$, $G : \underline{D} \to \underline{C}$.

Definition 1.13: Adjoint Functors
The functor F is a **left adjoint** of G, or G is a **right adjoint** of F, if there is a functorial isomorphism

$$\Theta : \text{Hom}_{\underline{D}}(F, -) \to \text{Hom}_{\underline{C}}(-, G)$$

where $\text{Hom}_{\underline{D}}(F, -) : \underline{C}^o \times \underline{D} \to \underline{\text{Set}}$ associates to (X, Y) the set $\text{Hom}_{\underline{D}}(F(X), Y)$; $\text{Hom}_{\underline{C}}(-, G) : \underline{C}^o \times \underline{D} \to \underline{\text{Set}}$ associates to (X, Y) the set $\text{Hom}_{\underline{C}}(X, G(Y))$.
If case $\underline{C}$ and $\underline{D}$ are pre-additive and the functors F and G are assumed to be additive, then we assume that $\Theta(X, Y)$ is an isomorphism of abelian groups.
The following sums up some basic facts concerning adjoint functors.

Properties 1.1
When the functor F is a left adjoint for G, then the following hold:

i. F commutes with coproducts, G commutes with products.

ii. If $\underline{C}$ and $\underline{D}$ are abelian and F and G are additive functors, then F is right exact and G is left exact.

iii. If for every object Y of $\underline{\mathcal{D}}$ there exists an injective object Q of $\underline{\mathcal{D}}$ and a monomorphism $Y \to Q$, then F is exact if and only if G preserves injectivity.

iv. If for every X of $\underline{\mathcal{C}}$ there exists a projective object P of $\underline{\mathcal{C}}$ together with an epimorphism $P \to X$ in $\underline{\mathcal{C}}$, then G is exact if and only if F preserves projectivity. Perhaps one of the most well-known pairs of adjoint functors appears in connection with module categories over associative rings R and T say. Consider the module categories R-mod and T-mod as well as R-mod-T, the category of left R-right-T-bimodules. For M in R-mod-T we may define the tensor-functor $M \otimes_T - : T\text{-mod} \to R\text{-mod}$ by viewing $M \otimes_T N$ for a left T-module N as a left R-module in the obvious way. It is easy to verify that $M \otimes_T -$ is a left adjoint of $\mathrm{Hom}_R(M, -) : R\text{-mod} \to S\text{-mod}$.

1.3 Separable Functors

The notion of **separable** functor has been introduced by M. Van den Bergh and the author; the concept has been applied to algebras and in particular graded algebras in [33]. Several other applications of ring theoretical nature stem from the paper by M. D. Rafael, an author among participants at a summer institute at Cortona. The separable functors are not absolutely necessary for the development of the theory in this work; nevertheless, we include a short presentation because they may be used in several applications and some research projects we cover.

Consider a covariant functor $F : \underline{\mathcal{D}} \to \underline{\mathcal{C}}$. The functor F is a **separable functor** if for all objects M, N in $\underline{\mathcal{D}}$ there are maps $\varphi^F_{M,N} : \mathrm{Hom}_{\underline{\mathcal{C}}}(F(M), F(N)) \to \mathrm{Hom}_{\underline{\mathcal{C}}}(M, N)$ satisfying the following properties:

SF.1 For $f \in \mathrm{Hom}_{\underline{\mathcal{D}}}(M, N)$, $\varphi^F_{M,N}(F(f)) = f$.

SF.2 For objects M', N' in $\underline{\mathcal{D}}$ and $f \in \mathrm{Hom}_{\underline{\mathcal{D}}}(M, M')$, $g \in \mathrm{Hom}_{\underline{\mathcal{D}}}(N, N')$, $f' \in \mathrm{Hom}_{\underline{\mathcal{C}}}(F(M), F(N))$, and $g' \in \mathrm{Hom}_{\underline{\mathcal{C}}}(F(M'), F(N'))$ such that the following diagram is commutative in $\underline{\mathcal{D}}$:

$$\begin{array}{ccc} F(M) & \xrightarrow{f'} & F(N) \\ {\scriptstyle F(f)}\downarrow & & \downarrow{\scriptstyle F(g)} \\ F(M') & \xrightarrow{g'} & F(N') \end{array}$$

then the following is commutative in $\underline{\mathcal{D}}$:

$$\begin{array}{ccc} M & \xrightarrow{\varphi^F_{M,N}(f')} & N \\ {\scriptstyle f}\downarrow & & \downarrow{\scriptstyle g} \\ M' & \xrightarrow{\varphi^F_{M',N'}(g')} & N' \end{array}$$

Observe that SF.1 holds if case F is a full faithful functor, that is, whenever for M, N in $\underline{\mathcal{D}}$ the map $\mathrm{Hom}_{\underline{\mathcal{D}}}(M, N) \to \mathrm{Hom}_{\underline{\mathcal{D}}}(F(M), F(N))$ is bijective.

Lemma 1.2

1. *An equivalence of categories is also a separable functor.*
2. *If $F : \underline{\mathcal{D}} \to \underline{\mathcal{C}}$ and $G : \underline{\mathcal{C}} \to \underline{\mathcal{B}}$ are separable functors, then GF is separable. Conversely, if GF is a separable functor, then F is a separable functor.*

Let us summarize some basic properties of separable functors in the following proposition.

Proposition 1.1

Let $F : \underline{\mathcal{D}} \to \underline{\mathcal{C}}$ be a separable functor and consider objects M and N in $\underline{\mathcal{D}}$.

i. *If $f \in \mathrm{Hom}_{\underline{\mathcal{D}}}(M, N)$ is such that $F(f)$ is a split map, then f itself is a split map.*
ii. *If $f \in \mathrm{Hom}_{\underline{\mathcal{D}}}(M, N)$ is such that $F(f)$ is co-split, that is to say that there exists a $u \in \mathrm{Hom}_{\underline{\mathcal{C}}}(F(N), F(M))$ such that $F(f) \circ u = 1_{F(N)}$, then f is itself co-split.*
iii. *Assume that both $\underline{\mathcal{D}}$ and $\underline{\mathcal{C}}$ are abelian categories. When F preserves epimorphisms, respectively monomorphisms, and $F(M)$ is projective in $\underline{\mathcal{C}}$, respectively injective in $\underline{\mathcal{C}}$, then M is projective in $\underline{\mathcal{D}}$, respectively injective in $\underline{\mathcal{D}}$.*
iv. *Assume that $\underline{\mathcal{D}}$ and $\underline{\mathcal{C}}$ are abelian categries. If $F(M)$ is a quasi-simple object, that is, every subobject splits off, and F preserves monomorphisms, then M is itself a quasi-simple object in $\mathcal{D}$.*

Proof

Statement iii follows from ii and iv follows i. The proof of ii is very similar to the proof of i, so it suffices to establish i. The assumption in i implies that there exists a map $u : F(N) \to F(M)$ such that $uF(f) = 1_{F(M)}$. Put $g = \varphi^F_{N,M}(u)$. Condition SF.2. then implies that $gf = 1_M$ because we do have a commutative diagram in $\underline{\mathcal{C}}$.

$$\begin{array}{ccc} F(M) & \xrightarrow{1_{F(M)}} & F(M) \\ {\scriptstyle F(f)}\downarrow & & \downarrow{\scriptstyle F(1_M)} \\ F(N) & \xrightarrow[u]{} & F(M) \end{array}$$

The claim follows. ☐

Corollary 1.2

Part i may be rephrased as a functorial version of Maschke's theorem (used frequently in the representation theory of groups).

We finish this section by pointing out that the terminology derives from the fact that the restriction of scalar functors associated to a ring morphism $C \to R$, where C is commutative and central in R, is a separable functor when R is C-separable. When R is also commutative, this agrees with the classical notion of a separable extension.

Chapter 2

Noncommutative Spaces

2.1 Small Categories, Posets, and Noncommutative Topologies

Throughout this section, $\mathcal{C}$ stands for a fixed small category, that is, a category having a class of objects that is a set. A category with exactly one object is a **monoid**; this is because we may view this as an object with a given monoid of endomorphisms. A group is then a monoid where all endomorphisms are automorphisms. By $\underline{O}$ we denote the zero-object category with a unique object O and a unique morphism: the identity of O. For any category $\mathcal{C}$ there exists a unique functor $\mathcal{C} \to \underline{O}$. A **terminal object** in $\mathcal{C}$ is an object, I say, such that for every object α of $\mathcal{C}$ there is a unique morphism $\alpha \to 1$ in $\mathcal{C}$. If $\mathcal{C}$ does not have a terminal object, we can adjoin one to $\mathcal{C}$ and obtain a category $\mathcal{C}_1$ with an obvious functor $\mathcal{C} \to \mathcal{C}_1$ taking an object α in $\mathcal{C}$ to α in $\mathcal{C}_1$.

To $\mathcal{C}$ we may associate the **opposite category** $\mathcal{C}^o$ having the same class of objects but with morphisms reversed. For an arbitrary category $\mathcal{D}$ a **$\mathcal{D}$-representation of $\mathcal{C}$** is just a functor $R : \mathcal{C} \to \mathcal{D}$; a $\mathcal{D}$-representation of $\mathcal{C}^o$ is called a **presheaf on $\mathcal{C}$ with values in $\mathcal{D}$**. Hence, a presheaf $P : \mathcal{C}^o \to \mathcal{D}$ is given by a family of objects $P(\alpha)$ in $\mathcal{D}$ such that for each $\mathcal{C}$-morphism $\alpha \to \beta$ we have $\mathcal{D}$-morphisms $\rho_\alpha^\beta : P(\beta) \to P(\alpha)$ such that to the identity $\alpha \to \alpha$ corresponds the identity $P(\alpha) \to P(\alpha)$, and to $\alpha \to \beta \to \gamma$ in $\mathcal{C}$ we correspond $\rho_\alpha^\gamma = \rho_\alpha^\beta \rho_\beta^\gamma$.

If $\mathcal{B}$ is another small category and given an arbitrary functor $F : \mathcal{B} \to \mathcal{C}$, we construct the left (and right) **comma category** as follows. For the objects of the right comma category (α, F) we take $\mathcal{C}$-morphisms $\alpha \to F\beta, \alpha \in \mathcal{C}, \beta \in \mathcal{B}$ and a morphism $(\alpha \to F\beta') \to (\alpha \to F\beta)$ in (α, F) as a $\mathcal{B}$-morphism $b : \beta' \to \beta$ making the following diagram commutative:

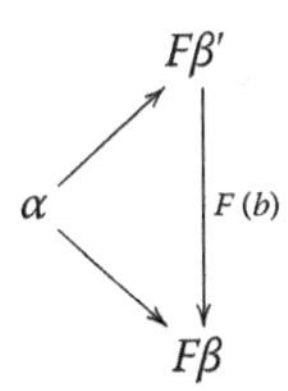

The left comma category (F, α) is defined likewise, using for the objects the $\mathcal{C}$-morphisms $F\beta \to \alpha$, and so forth.

Any $\mathcal{C}$-morphism $a : \alpha' \to \alpha$ induces functors:

$$(a, F) : (\alpha, F) \to (\alpha', F), (\alpha \to F\beta) \mapsto (\alpha \to F\beta)a$$
$$(F, a) : (F, \alpha') \to (F, \alpha)$$

A type of small category often considered is a poset. A **poset, or partially ordered set**, is just a set with a partial ordering: $\leq$. If Λ is a poset, then we shall write $\underline{\Lambda}$ for the category having as objects the elements $\lambda \in \Lambda$, and $\hom_{\underline{\Lambda}}(\lambda, \mu)$ consists of the unique arrow $\lambda \to \mu$ when $\lambda \leq \mu$, or it is empty. The categories $\underline{\Lambda}$ are examples of **delta categories**, that is, small categories in which endomorphisms are necessarily the identity morphisms and $\hom(\sigma, \tau) \neq \emptyset$ for $\sigma \neq \tau$ implies $\hom(\tau, \sigma) = \emptyset$ (maps are one-way and no loops).

We define **$\mathcal{D}$-representations of Λ, presheaves on Λ with values in $\mathcal{D}$, and comma categories for $\lambda \in \Lambda, \ldots$** by taking the corresponding definitions for $\underline{\Lambda}$.

The mother of all posets is the set of natural numbers with its usual ordering. For $n \in \mathbb{N}$ we let $[n]$ denote the linearly ordered set $0 < \cdots < n$ viewed as a category (as for posets). A $\mathcal{C}$-representation of $[n]$ is called an n**-simplex**; if $\sigma : [n] \to \underline{\mathcal{C}}$ is a (covariant) functor, then we say that σ is an n-simplex or $\dim\sigma = n$. We denote the $\mathcal{C}$-morphism $\sigma(r \to s)$ by σ^{rs}, for $r \leq s$ in $[n]$. Zero simplices are functors $[O] \to \mathcal{C}$; these may be viewed as the elements of $\underline{\mathcal{C}}$, up to a harmless "abuse of language." For $n > 0$, an n-simplex σ is completely determined by the n-tuple $(\sigma^{01}, \sigma^{12}, \ldots, \sigma^{p-1,p})$; therefore, it is unambiguous to write $\sigma = (\sigma^{01}, \sigma^{12}, \ldots)$.

When $\mathcal{C}$ is $\underline{\Lambda}$ for some poset Λ, then an n-simplex σ is completely determined by the ordered list of elements, called vertices, $\sigma(0), \ldots, \sigma(n)$, because any σ^{rs} is then necessarily the unique $\underline{\Lambda}$-morphism $\sigma(r) \to \sigma(s)$ corresponding to $r \leq s$.

If τ is an n-simplex and σ is an m-simplex such that $\tau(n) = \sigma(0)$, then we can form the **cup-product** $\tau \vee \sigma$, which is the $(n+m)$-simplex given by:

$$\begin{aligned} (\tau \vee \sigma)^{r,r+1} &= \tau^{r,r+1} && \text{when } r < n \\ (\tau \vee \sigma)^{r,r+1} &= \sigma^{r-n,r-n+1} && \text{when } n \leq r. \end{aligned}$$

For $n > 0$ and $o \leq r \leq n$ we define a functor

$$\partial_r : [n-1] \to [n], \quad \begin{cases} \partial_r(s) = s & \text{if } s < r \\ \partial_r(s) = s + 1 & \text{if } s \geq r. \end{cases}$$

A given n-simplex σ has an r**-face** defined by the composition of the functors σ and ∂_r.

For $r \geq s$ one easily computes $\partial_r \partial_s = \partial_s \partial_{r+1}$, where composition is here written in the arrow-order (i.e., not the usual way of writing a composition of maps). Hence:

$$\sigma_r = \begin{cases} (\sigma^{12}, \ldots, \sigma^{n-1,n}) & \text{if } r = 0 \\ (\sigma^{01}, \ldots, \sigma^{r-1,r+1}, \ldots, \sigma^{n-1,n}) & \text{if } 0 < r < n \\ (\sigma^{01}, \ldots, \sigma^{n-1,n-1}) & \text{if } r = n \end{cases}$$

In case $\mathcal{C} = \underline{\Lambda}$, then the faces are distinct but that need not be true in general for arbitrary $\mathcal{C}$.

The ∂_r are called **face operators**. The collection of n-simplices and the face operators connecting them are called the **simplicial complex**, $\Sigma(\mathcal{C})$. It will be useful for introducing homology theories in a formal way.

A **zero element**, denoted by 0, of the poset Λ is one for which $0 \leq \lambda$ for all $\lambda \in \Lambda$; clearly, if a zero element exists it is unique. A **unit element**, denoted by 1, of the poset Λ is one for which $\lambda \leq 1$ for all $\lambda \in \Lambda$; if a unit element exists then it is unique. A poset Λ with 0 and 1 is said to be a **lattice** if for any two elements λ and μ in Λ there exists a maximum $\lambda \vee \mu \in \Lambda$ and a minimum $\lambda \wedge \mu \in \Lambda$.

A lattice is said to be $\vee$**-complete** if for any family $\{\lambda_\alpha, \alpha \in \mathcal{A}\}$ in Λ the maximum $\vee_{\alpha \in \mathcal{A}} \lambda_\alpha$ exists in Λ. The lattice Λ is **complete** if it is both $\vee$-complete and $\wedge$-complete. For the general theory of lattices we refer to [6][7]. Now let us introduce the notion of **cover** in an arbitrary poset Λ.

Definition 2.1
We say that $\lambda \in \Lambda$ is **covered** by $\{\lambda_\alpha, \alpha \in \mathcal{A}\}$ with $\lambda_\alpha \in \Lambda$ for all $\alpha \in \mathcal{A}$, if $\lambda_\alpha \leq \lambda$ for all $\alpha \in \mathcal{A}$ and if $\lambda_\alpha \leq \mu$ for all $\alpha \in \mathcal{A}$ for some $\mu \in \Lambda$ then $\lambda \leq \mu$; in this case we also say that $\{\lambda_\alpha, \alpha \in \mathcal{A}\}$ is a cover for λ. If $\mathcal{A}$ is finite, then $\{\lambda_\alpha, \alpha \in \mathcal{A}\}$ is said to be a **finite cover** for $\lambda \in \Lambda$.

Example 2.1

i. If Λ is a lattice, then $\{\lambda_1, \ldots, \lambda_n\}$ is a finite cover for $\lambda \in \Lambda$ exactly when $\lambda = \lambda_1 \vee \ldots \vee \lambda_n$; in a complete lattice this holds for arbitrary covers.

ii. If Λ is a distributive lattice, then a given finite cover $\lambda = \lambda_1 \vee \ldots \vee \lambda_n$ induces for every $\tau \leq \lambda$ a cover: $\tau = (\tau \wedge \lambda_1) \vee \ldots \vee (\tau \wedge \lambda_n)$; the latter is called the **induced cover** for $\tau \in \lambda$.

iii. If Λ is a poset with 1, then a **global cover** is a set $\{\lambda_\alpha, \alpha \in \mathcal{A}\}$ such that $\mu \geq \lambda_\alpha$ for all $\alpha \in \mathcal{A}$ entails $\mu = 1$. In particular, if Λ is a distributive lattice with 0 and 1, then a finite global cover is a set $\{\lambda_1, \ldots, \lambda\}$ such that $1 = \lambda_1 \vee \ldots \vee \lambda_n$ and every $\tau \in \Lambda$ then allows a cover $\{\tau \wedge \lambda_1, \ldots, \tau \wedge \lambda_n\}$ induced by a global cover.

iv. A cover $\{\lambda_1, \ldots, \lambda\}$ of λ is **reduced** when it is $\vee$-independent in the lattice Λ. In a semi-atomic lattice with 0 and 1 that is upper continuous and has the property that 1 can be written as a finite join of atoms of Λ, there always exists a reduced global cover.

2.1.1 Sheaves over Posets

In this section we restrict attention to a category $\underline{\mathcal{D}}$, the objects of which are sets; hence, morphisms in $\underline{\mathcal{D}}$ are in particular set maps.

A presheaf $P : (\underline{\Lambda})^o \to \underline{\mathcal{D}}$ is **separated** if for every cover $\{\lambda_\alpha, \alpha \in \mathcal{A}\}$ of λ in Λ and every $x, y \in P(\lambda)$ such that for all $\alpha \in \mathcal{A}$, $\rho^{\lambda}_{\lambda_\alpha}(x) = \rho^{\lambda}_{\lambda_\alpha}(y)$, we must have $x = y$. In case no covers exist, then every presheaf is separated. A separated presheaf is a **sheaf** on Λ (with values in $\underline{\mathcal{D}}$), if for every finite cover $\{\lambda_i, i\}$ of λ and given

$x_i \in P(\lambda_i)$ such $\rho_\mu^{\lambda_i}(x_i) = \rho_\mu^{\lambda_j}(x_j)$ for every $\mu \leq \lambda_i, \mu \leq \lambda_j$, there exists an $x \in P(\lambda)$ such that for all i, $\rho_{\lambda_i}^{\lambda}(x) = x_i$.

In order to introduce stalks of presheaves or sheaves, we first introduce a so-called limit poset $C(\Lambda)$ associated to any given poset Λ.

2.1.2 Directed Subsets and the Limit Poset

A subset $X \subset \Lambda$ is said to be **directed** if for every x, y in X there exists a $z \in X$ such that $z \leq x$ and $z \leq y$. Let $\mathcal{D}(\Lambda)$ be the set of directed subsets of Λ. For A and B in $\mathcal{D}(\Lambda)$ we say that A **is equivalent to** B, written $A \sim B$, if and only if the following conditions are satisfied:

i. For $a \in \Lambda$ there exists $a' \in A, a' \leq a$ and $b, b' \in B$ such that: $b \leq a' \leq b'$.

ii. For $b \in B$, there exists $b' \in B, b' \in b$ and $a, a' \in A$ such that: $a \leq b' \leq a'$.

By $[A]$ we denote the $\sim$-equivalence class of A in $\mathcal{D}(\Lambda)$.

We let $C(\Lambda)$ be the set of classes of directed subsets of Λ. A directed set $X \subset \Lambda$ is said to be a **filter** in Λ if $x \leq y$ with $x \in X$ entails $y \in X$. To an arbitrary directed set Y in Λ we associate a filter $\overline{Y}$ as follows:

$$\overline{Y} = \{\lambda \in \Lambda, \text{there exists an } x \in Y \text{ such that } x \leq \lambda\}.$$

For A, B in $\mathcal{D}(\Lambda)$ we say that $A \leq B$ if and only if:

i. For $a \in A$ there is an $a' \in A, a' \leq a$, such that $a' \leq b$ for some $b \in B$.

ii. For $b \in B$, there is an $a \in A$ such that $a \leq b$.

Lemma 2.1

With conventions and notation as above:

1. *For A, B in $\mathcal{D}(\Lambda)$, $A \leq B$ if and only if $\overline{B} \subset \overline{A}$.*
2. *For A, B in $\mathcal{D}(\Lambda)$, $A \sim B$ if and only if $A \leq B$ and $B \leq A$, if and only if $\overline{A} = \overline{B}$.*
3. *The set $C(\Lambda)$ with $\leq$ induced from $\mathcal{D}(\Lambda)$ is a poset such that the canonical map $\Lambda \to C(\Lambda)$, $\lambda \mapsto [\{\lambda\}]$, is a poset monomorphism.*
4. *If Λ is a lattice with respect to $\wedge, \vee$ then these operations induce a lattice structure on $C(\Lambda)$ such that the canonical poset map $\Lambda \to C(\Lambda)$ is a lattice monomorphism.*
5. *If Λ is a lattice, then the following properties of Λ transfer to the lattice $C(\Lambda)$:*
   i. Λ has 0 and 1.
   ii. Λ is modular.
   iii. Λ is complete.

iv. Λ *is distributive.*

v. Λ *is Brouwerian, that is, for* $\lambda, \mu \in \Lambda$ *the set* $\{x \in \Lambda, \lambda \wedge x \leq \mu\}$ *has a largest element then denoted by* $\mu : \lambda$.

Proof

1, 2, and 3 are easy enough, and we leave these as exercises.

4. For A and B in $\mathcal{D}(\Lambda)$ define:

$$A \dot{\wedge} B = \{a \wedge b, a \in A, b \in B\},\ A \dot{\vee} B = \{a \vee b, a \in A, b \in B\}$$

Since $\wedge$ and $\vee$ are bicontinuous with respect to $\leq$, it is immediately clear that $A \dot{\wedge} B$ as well as $A \dot{\vee} B$ is in $\mathcal{D}(\Lambda)$. If $A \sim A'$ and $B \sim B'$ in $\mathcal{D}(\Lambda)$, then $A \dot{\wedge} B \sim A' \dot{\wedge} B'$ and $A \dot{\vee} B \sim A' \dot{\vee} B'$. So if we define $[A] \wedge [B] = [A \dot{\wedge} B]$, $[A] \vee [B] = [A \dot{\vee} B]$ in $C(\Lambda)$, then we obtain a well-defined lattice structure on $C(\Lambda)$ making the canonical poset map $\Lambda \to C(\Lambda)$, $\lambda \mapsto [\{\lambda\}]$ into a lattice monomorphism, as is easily checked.

5. Straightforward verification of the lattice properties considered. □

Exercise 2.1

1. *For any set S we let $L(S)$ be the poset of all subsets of S partially ordered by inclusion. A* ***lattice presentation*** *of a poset Λ is just a poset map $\pi : \Lambda \to L(S)$ for some set S; observe that $L(S)$ is a lattice. Verify that it is always possible to present a poset Λ by $\pi : \Lambda \to L(\Lambda)$, $\lambda \mapsto \{\mu \in \Lambda, \mu \leq \lambda\}$. Provide examples to show that π need not be injective or surjective in general.*

2. *For any poset Λ a subset of the form $\{\mu \in \Lambda, \mu \leq \lambda\}$ is called an* ***interval*** *of Λ. For a lattice Λ we have a lattice isomorphism $\Lambda \simeq I(\Lambda)$ described by $\lambda \leftrightarrow \{\mu \in \Lambda, \mu \leq \lambda\}$, where $I(\Lambda)$ is the lattice of intervals in Λ.*

3. *If $\pi : \Lambda \to L(S)$ is a lattice presentation of the poset Λ, then it induces a poset map $C(\pi) : C(\Lambda) \to CL(S)$ fitting in the following commutative diagram of poset maps:*

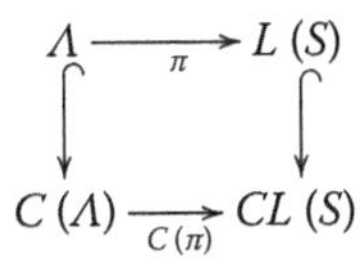

Verify also that $C(\pi)$ actually defines a lattice presentation of $C(\Lambda)$, $C(\Lambda) \to L(CL(S))$. Note that we may identify $CL(S)$ and $I(CL(S))$ in view of Lemma 2.1(4).

4. *Can you describe $C(\Lambda)$ in case Λ is the lattice of open sets*

   a. *for the real topology on $\mathbb{R}^n$*

   b. *for the Zariski topology on $\mathbb{R}^n$*

5. *If* Λ *is the lattice* Open(X) *consisting of open sets of a topological space* X, τ, *can you express the Hausdorff, respectively* T_1*-property, in terms of the lattice morphism* $\Lambda \to C(\Lambda)$*?*

6. *If* Λ_1, Λ_2 *are posets, respectively lattices, is it possible to define a product poset, respectively product lattice,* $\Lambda_1 \times \Lambda_2$*? Can you relate* $C(\Lambda_1) \times C(\Lambda_2)$ *and* $C(\Lambda_1 \times \Lambda_2)$*? In case* $\Lambda_1 = \text{Open}(X)$, $\Lambda_2 = \text{Open}(Y)$, *is* $\Lambda_1 \times \Lambda_2$ *equal to Open*$(X \times Y)$ *where* $Y \times Y$ *is equipped with the product topology? Relate* $C(\text{Open}(X \times Y))$ *and* $C(\Lambda_1)$, $C(\Lambda_2)$. *In case* X *and* Y *are varieties with their respective Zariski topologies and* $X \times Y$ *is viewed with the Zariski topology (i.e., this need not be the product topology of the product variety* $X \times Y$*), is it true that* $C(\text{Open}(X \times Y)) = C(\text{Open}_{\text{Zar}}(X \times Y))$*? Is the relation between the lattices* $C(\text{Open}_{\text{Zar}}(X)) \times C(\text{Open}_{\text{Zar}}(Y))$ *and* $C(\text{Open}_{\text{Zar}}(X \times Y))$ *essentially different from the relation between the topologies themselves?*

2.1.3 Poset Dynamics

When deformations of algebras are introduced, even when using rather general abstract methods such as diagram algebras in the sense of M. Gerstenhaber, there is always some algebra or ring of formal power series in the picture. This is somewhat unsatisfactory because it is really not asking too much to hope that the use of some sheaf-like theory over a suitably general underlying "space" (topology, lattice, poset, diagram, etc.) should allow a visualization of the deformation action and an interpretation in terms of infinitesimal-like phenomena worded in terms of sheaves in suitable categories. Motivated by a question concerning the possibility of developing a dynamic version of the use of causets (posets with respect to a causality order) in the theory of quantum gravity, we propose a notion of **poset dynamics**. This structure may be viewed as an interesting toy; that is, a generic exercise arises when trying to apply any derived structural result in the usual, that is to say static, theory to the dynamic situation.

Let T be a totally ordered poset. A **poset** T**-dynamics** consists of a class of posets $\{P_t, t \in T\}$ together with poset maps: $\varphi_{tt'} : P_t \to P_{t'}$ for every $t \le t'$ in T, satisfying the following conditions:

DP.1

For each $t \in T$, $\varphi_{tt} = I_{P_t}$, the identity of P_t.

DP.2

For each triple $t \le t' \le t''$ in T we have: $\varphi_{tt'}\varphi_{t't''} = \varphi_{tt''}$ (where composition is written in the arrow-order).

DP.3

For any $t \in T$ and subset $\mathcal{F} \subset T$ such that for every $f \in \mathcal{F}$ we have $t \le f$, for every nontrivial $x < y$ in P_t (i.e., $x \ne 0, y \ne 1$) such that for all $f \in \mathcal{F}$, $\varphi_{tf}(x) < z_f < \varphi_{tf}(y)$ for some $z_f \in P_f$, there exists a $z \in P_t$ with $x < z < y$ such that $\varphi_{tf}(x) < \varphi_{tf}(z) \le z_f < \varphi_{tf}(y)$.

DP.4
For every $t \in T$ and nontrivial $x < z < y$ in P_t (i.e., x and y not 0 and 1), there exists a $t_1 \in T, t < t_1$, such that for all $t' \in [t, t_1[$ we have $\varphi_{tt'}(x) < \varphi_{tt'}(z) < \varphi_{tt'}(y)$ (in other words strict $<$ has a future).

DP.5
For every $t \in T$ and $x < z < y$ in P_t there exists a $t_1 \in T$, $t_1 < t$, such that for all $t' \in]t_1, t]$ such that $x', y' \in P_{t'}$ with $\varphi_{t't}(x') = x, \varphi_{t't}(y') = y$ exist, we have $x' < z' < y'$ in $P_{t'}$ with $\varphi_{t't}(z') = z$ (in other words strict $<$ has a past for elements with a past).

Note the discrepancy between **DP.3** and **DP.5**; in **DP.3** we reach P_t but the condition $\varphi_{tf}(z) \leq z_f$ is weaker than $\varphi_{t't}(z') = z$ in **DP.5**, which is only reached on $P_{t'}$. Under conditions ensuring that the set of z having the property in **DP.3** has maximal elements (e.g., $\vee$-complete, Noetherian posets, etc.), then we may by iteration in **DP.3** arrive at a z such that $\varphi_{tf}(z) = z_f$ and then **DP.3** implies **DP.5**. In that case Imφ_{tf} is **convex** in Λ_f, $(\varphi_{tf}([x, y]) = [\varphi_{tf}(x), \varphi_{tf}(y)])$.

The remainder of this section is devoted to the introduction of the notion of dimension of a poset. This in turn may be used to define the Krull dimension of an abelian category. Both concepts will have a different meaning in applications, comparable to the difference between the notion of dimension of a topological space and the dimension of a variety or of a linear space. The Krull dimension of a poset was introduced by P. Gabriel and R. Rentschler [11] for ordinal numbers; a generalization for "higher" numbers is obtained later by G. Krause.

Let $\Lambda, \leq$ be a poset. If $\lambda \leq \mu$ in Λ, then we write $[\lambda, \mu]$ for the closed interval $\{\alpha \in \Lambda, \lambda \leq \alpha \leq \mu\}$. The set of intervals is denoted by $\mathcal{I}(\Lambda)$. We may define on $\mathcal{I}(\Lambda)$ the following filtration by using transfinite recurrence

$$\mathcal{I}_{-1}(\Lambda) = \{[\lambda, \mu], \lambda = \mu\}$$
$$\mathcal{I}_0(\Lambda) = \{[\lambda, \mu] = \mathcal{I}(\Lambda), [\lambda, \mu] \text{ is an Artinian poset}\}.$$

(Artinian means that the descending chain condition holds with respect to subposets, or equivalently every nonempty family in it has a minimal element). Assuming that $\mathcal{I}_\beta(\Lambda)$ has already been defined for all $\beta < \alpha$, then we define $\mathcal{I}_\alpha(\Lambda)$ as follows: $\mathcal{I}_\alpha(\Lambda) = \{[\lambda, \mu]$, for every sequence $\lambda \geq \lambda_1 \geq \ldots \lambda_n \geq \ldots \mu$ there is an $n \in \mathbb{N}$ such that $[\lambda_{i+1}, \lambda_i] \in \cup_{\beta<\alpha}\mathcal{I}_\beta(\Lambda)$ for all $i \geq n\}$.

We have obtained the ascending chain:

$$\mathcal{I}_{-1}(\Lambda) \subset \mathcal{I}_0(\Lambda) \subset \cdots \subset \mathcal{I}_\alpha(\Lambda) \subset \cdots$$

If there exists an ordinal α such that $\mathcal{I}(\Lambda) = \mathcal{I}_\alpha(\Lambda)$, then Λ is said to have **Krull dimension**. The smallest ordinal α such that $\mathcal{I}(\Lambda) = \mathcal{I}_\alpha(\Lambda)$ is called the **Krull dimension of** Λ and we denote it by $K \dim \Lambda$.

Lemma 2.2
If $f : \Lambda \to \Gamma$ is a strictly increasing map of posets, then Λ has Krull dimension if Γ has Krull dimension and $K \dim \Lambda \leq K \dim \Gamma$.

Proof
See [33]. □

Lemma 2.3

Let Λ and Γ be posets having Krull dimension. Consider the product poset $\Lambda \times \Gamma$ with the product ordering. Then $\Lambda \times \Gamma$ has Krull dimension and $K \dim(\Lambda \times \Gamma) = \sup\{K \dim \Lambda, K \dim \Gamma\}$.

Proof

See [33]. □

Now let $\underline{A}$ be an arbitrary abelian category and M an object of $\underline{A}$. Let $L(M)$ be the class of all subobjects of M in $\underline{A}$ ordered by "inclusion"; in fact $L(M)$ is a (big) modular lattice. If $M = 0$, then we put $K \dim M = -1$; if α is an ordinal and $K \dim M$ is not smaller than α, then we put $K \dim M = \alpha$ provided there is no infinite chain $M \supset M_0 \supset M_1 \supset \ldots \supset \ldots$ of subobjects M_i of M in $\underline{A}$ such that for $i \geq 1$, $K \dim M_{i-1}/M_i$ is not smaller than α. An object M of $\mathcal{A}$ having $K \dim M = \alpha$ is said to be α-**critical** if $K \dim(M/M') < \alpha$ for every nonzero subobject M' of M in $\underline{A}$. For example, M is o-critical if and only if M is a simple object of $\underline{A}$. Obviously, a nonzero subobject of an α-critical object is again α-critical.

Lemma 2.4

Let N be a subobject of M in $\underline{A}$; then $K \dim M \leq \sup\{K \dim N, K \dim M/N\}$*; equality holds provided either side exists.*

Proof

See Lemma B.1.2 in reference [33] □

Lemma 2.5

If an object M of $\underline{A}$ has Krull dimension, then it contains a critical subobject.

Proof

See [11] or [33]. □

Lemma 2.6

Every Noetherian object of $\underline{A}$ has Krull dimension.

Lemma 2.7

Suppose that $\underline{A}$ is an abelian category allowing arbitrary coproducts. If M, an object of $\underline{A}$ has Krull dimension, then M cannot contain an infinite direct sum (coproduct) of subobjects.

Lemma 2.8

If $\underline{A}$ is as before, let M be an object of $\underline{A}$ having Krull dimension. Put $\alpha = \sup\{K \dim(M/N) + 1$*, N an essential subobject of $M\}$, where N is essential in the sense that for any subobject P of M we must have $P \cap N \neq 0$. Then* $\mathrm{Kdim} M \leq \alpha$.

Lemma 2.9
Let M be a Noetherian object of $\underline{A}$. There exists a composition series: $M \supset M_1 \supset \ldots \supset M_n = 0$ such that M_{i-1}/M_i is a critical object for each $1 \leq i \leq n$. Moreover, if $\alpha_i = K \dim(M_{i-1}/M_i)$, then $K \dim M = \sup\{\alpha_i, 1 \leq i \leq n\}$.

Lemma 2.10
Let U be a generator for the category $\underline{A}$ and let M be an object of $\underline{A}$. If U and M both have Krull dimension, then $K \dim M \leq K \dim U$.

In the situation of the foregoing lemma we may take $K \dim \underline{A} = K \dim U$; this situation is clarified further when we introduce the notion of Gabriel dimension; compare Section 2.6.

2.2 The Topology of Virtual Opens and Its Commutative Shadow

In [46] we introduced a **noncommutative topology** as a poset Λ with operations $\wedge$ and $\vee$ satisfying the following set of axioms: (note that we assume $0, 1 \in \Lambda$):

A.1 For $x, y \in \Lambda$, $x \wedge y \leq y$.

A.2 For $x \in \Lambda$, $x \wedge 1 = 1 \wedge x = x$ and $x \wedge 0 = 0 \wedge x = 0$; moreover $x \wedge \cdots \wedge x = 0$ if and only if $x = 0$.

A.3 For $x, y, z \in \Lambda$, $x \wedge y \wedge z = (x \wedge y) \wedge z = x \wedge (y \wedge z)$.

A.4 For $a \leq b$ in Λ and $x, y \in \Lambda$ we obtain $x \wedge a \leq x \wedge b$, $a \wedge y \leq b \wedge y$ (bicontinuity).

A.5 For $x, y \in \Lambda$, $y \leq x \vee y$.

A.6 For $x \in \Lambda$ we have: $1 \vee x = x \vee 1 = 1$, $x \vee 0 = 0 \vee x = x$; moreover $x \vee \ldots \vee x = 1$ if and only if $x = 1$.

A.7 For $x, y, z \in \Lambda$, $x \vee y \vee z = (x \vee y) \vee z = x \vee (y \vee z)$.

A.8 For $a \leq b$ in Λ and $x, y \in \Lambda$ we obtain $x \vee a \leq x \vee b$, $a \vee y \leq b \vee y$.

A.9 Let $\mathrm{id}_\wedge(\Lambda)$ be the set of $\wedge$-idempotent elements of Λ, that is, $\lambda \in i_\wedge(\Lambda)$ if and only if $\lambda \wedge \lambda = \lambda$. For $x \in \mathrm{id}_\wedge(\Lambda)$ and $x \leq z$ in Λ we have $x \vee (x \wedge z) \leq (x \vee x) \wedge z$, $x \vee (z \wedge x) \leq (x \vee z) \wedge x$.

A.10 For $x \in \Lambda$ and $\lambda_1, \ldots, \lambda_n \in \Lambda$ such that $1 = \lambda_1 \vee \cdots \vee \lambda_n$ we have $(x \wedge \lambda_1) \vee \cdots \vee (x \wedge \lambda_n) = x$.

When in the above axioms A.4 and A.8 the statements are restricted to operations on the left, then Λ is said to be a **left topology**; note that A.10 as it has been formulated above only states that global covers induce covers on the left, that is, by taking $\wedge$ from the left. In case one wants to define a **right topology**, the axiom A.10 also has to be modified so that global covers induce covers via $\wedge$ on the right. A left and right topology is then a noncommutative topology where A.10 holds for inducing covers both from the left or from the right.

2.2.1 Properties

1. From $A_1, \ldots, A_4$, it follows that $x \wedge y, y \wedge x \leq x, y$.
2. If $x \in \mathrm{id}_\wedge(\Lambda)$ and $x \leq y$, then $x \wedge y = y \wedge x = x$.
3. For $x, y \in \mathrm{id}_\wedge(\Lambda)$, $x \wedge y = y \wedge x$ yields $x \wedge y \in \mathrm{id}_\wedge(\Lambda)$. Moreover, if $x \wedge y$ and $y \wedge x$ are in $i_\wedge(\Lambda)$, then $x \wedge y = y \wedge x$.
4. If Λ satisfies $A.1, \ldots, A.9$, then $\mathrm{id}_\wedge(\Lambda) \subset \mathrm{id}_\vee(\Lambda)$.
5. For $x \in \mathrm{id}_\wedge(\Lambda)$ and $x \leq z$ we obtain $x \vee (y \wedge z) \leq ((x \vee y) \wedge z)^{\vee 2} (x \vee (y \wedge z))^{\wedge 2} \leq (x \vee y) \wedge z$; where $\vee 2$ and $\wedge 2$ are exponent notation with respect to $\vee$ and $\wedge$, respectively.
6. For $x, y, z \in \Lambda$ with $x \in \mathrm{id}_\wedge(\Lambda)$ we obtain

$$x \wedge (y \vee z) \geq ((x \wedge y) \vee (x \wedge z))^{\wedge 2}$$
$$((x \vee y) \wedge (x \vee z))^{\vee 2} \geq x \vee (y \wedge z).$$

We say that $x \leq \lambda$ is **focused** on x if $\lambda \wedge x = x \wedge \lambda = x$; clearly for idempotent $x \in \mathrm{id}_\wedge(\Lambda)$ the relation $x \leq \lambda$ is focused. We say that Λ satisfies **FDI (focused distributive identity)** if for a focused relation $x \leq \lambda$ with $\lambda = \lambda_1 \vee \lambda_2$ we have $x = (x \wedge \lambda_1) \vee (x \wedge \lambda_2)$.

The term *noncommutative topology* has appeared in the literature before, but in fact the structures considered would have to be called "commutative" in our philosophy because the essential noncommutative aspect, allowing $A \wedge A \neq A$ for some A, is never present.

In a noncommutative topology Λ we may consider a subset $\mathcal{T}(\Lambda)$ consisting of finite length bracketed (if Λ allows infinite $\vee$ then this can be allowed in constructing $\mathcal{T}(\Lambda)$) expressions involving $\wedge$, $\vee$ and elements of $i_\wedge(\Lambda)$. Then $\mathcal{T}(\Lambda)$ again satisfies A.1, ..., A.10, so we call it the noncommutative topology generated by $\mathrm{id}_\wedge(\Lambda)$.

An idempotent $\lambda \in i_\wedge(\lambda)$ is called **contracting** for $\lambda \leq \lambda_1 \vee \lambda_2$; we have that $\lambda = (\lambda \wedge \lambda_1) \vee (\Lambda \wedge \lambda_2)$. For example, when Λ satisfies FDI, then every idempotent is contracting (so this will hold for the examples in noncommutative geometry). A noncommutative topology $T(\Lambda)$ is said to be a topology of **virtual opens** if it satisfies the following conditions:

VOT.1 The operation $\vee$ is commutative.

VOT.2 For every family $\mathcal{F}$ of elements in $\mathcal{T}(\Lambda)$, $\vee\mathcal{F} = \vee\{\lambda \in \mathcal{F}\}$ exists in Λ, where $\vee\mathcal{F}$ is characterized by the properties $\lambda \leq \vee\mathcal{F}$ for all $\lambda \in \mathcal{F}$ and if $\lambda \leq \mu$ for all $\lambda \in \mathcal{F}$, then $\vee\mathcal{F} \leq \mu$.

VOT.3 $\mathcal{T}(\Lambda)$ is generated as a noncommutative topology by a set of contracting idempotents, that is, every $\lambda \in \mathcal{T}(\Lambda)$ can be written as a finite length bracketed expression involving $\vee$, $\wedge$ and contracting idempotents of Λ.

Observe that if $\mathcal{T}$ is generated by a set S of contracting idempotents, we do have $S \subset \mathrm{id}_\wedge(\mathcal{T})$ but equality need not hold. Easy examples of this situation are obtained by considering any classical distributive lattice, where indeed every element is a contracting idempotent.

Allowing arbitrary $\vee\mathcal{F}$ in the definition of bracketed expressions, that is, only demanding finiteness with respect to the $\wedge$ operation, and also in the definition of contracting idempotents, we obtain the definition of a **$\vee$-complete topology of virtual opens**. We say that $\mathcal{T}$ is **$\vee$-compact** if for every $\mathcal{F}$ there exists a finite $\mathcal{F}'$ in $\mathcal{F}$ such that $\vee\mathcal{F} = \vee\mathcal{F}'$. Since the opposite poset $\mathcal{T}^{\mathrm{op}}$ is not a skew-topology, the specification *to* $\vee$ in the definition of $\vee$-compact is not superfluous. Unless otherwise stated, we now let Λ be a topology of virtual opens. We have observed that $\mathrm{id}_\wedge(\Lambda)$ does not inherit a similar structure with respect to $\wedge$. For $\sigma, \tau \in \mathrm{id}_\wedge(\Lambda)$ define $\sigma \underset{\cdot}{\wedge} \tau$ by taking $\vee\{\gamma \in \mathrm{id}_\wedge(\Lambda), \gamma \leq \sigma \wedge \tau\}$. We list some elementary properties in the following lemma.

Lemma 2.11
With notation and conventions as above:

i. *If* $\sigma \in \mathrm{id}_\wedge(\Lambda)$, *then* $\sigma \underset{\cdot}{\wedge} \sigma = \sigma$ *(and conversely).*

ii. *If* $\sigma \in \mathrm{id}_\wedge(\Lambda)$ *and* $\sigma \underset{\cdot}{\wedge} \sigma \underset{\cdot}{\wedge} \ldots \underset{\cdot}{\wedge} \sigma = 0$, *then* $\sigma = 0$.

iii. *If* $\tau \leq \gamma$ *in* Λ, *then* $\sigma \underset{\cdot}{\wedge} \tau \leq \sigma \underset{\cdot}{\wedge} \gamma$ *for* $\sigma \in \Lambda$; *similarly* $\tau \underset{\cdot}{\wedge} \sigma \leq \gamma \underset{\cdot}{\wedge} \sigma$.

iv. *For* $\tau, \gamma \in \Lambda$, $\tau \underset{\cdot}{\wedge} \gamma \in \mathrm{id}_\wedge(\Lambda)$.

v. *For* $\sigma, \tau, \gamma \in \Lambda$: $\sigma \underset{\cdot}{\wedge} (\tau \underset{\cdot}{\wedge} \gamma) = (\sigma \underset{\cdot}{\wedge} \tau) \underset{\cdot}{\wedge} \gamma$.

vi. *The axiom A.9 holds for* $\mathrm{id}_\wedge(\Lambda), \underset{\cdot}{\wedge}$.

Proof

i. Easy enough.

ii. Clear because for $\sigma \in \mathrm{id}_\wedge(\Lambda)$, $\sigma \underset{\cdot}{\wedge} \sigma = \sigma$.

iii. Consider σ, τ, γ in $\mathrm{id}_\wedge(\Lambda)$ with $\tau \leq \gamma$. Then $\sigma \underset{\cdot}{\wedge} \tau$ is $\wedge$-idempotent (see iv) and $\sigma \underset{\cdot}{\wedge} \tau \leq \sigma \wedge \tau \leq \sigma \wedge \gamma$, hence $\sigma \underset{\cdot}{\wedge} \leq \sigma \underset{\cdot}{\wedge} \gamma$. The right symmetric version follows in a similar way.

iv. We have $\sigma \underset{\cdot}{\wedge} \gamma = \vee\{\tau, \tau \in \mathrm{id}_\wedge(\Lambda), \tau \leq \sigma \wedge \gamma\}$. Now look at $(\sigma \underset{\cdot}{\wedge} \gamma) \wedge (\sigma \underset{\cdot}{\wedge} \gamma)$ and observe that this is larger than $\tau \wedge \tau$ for every idempotent $\tau \leq \sigma \wedge \gamma$, that is, $(\sigma \underset{\cdot}{\wedge} \wedge (\sigma \underset{\cdot}{\wedge} \gamma) \geq \vee\{\tau \in \mathrm{id}_\wedge(\Lambda), \tau \leq \sigma \wedge \gamma\} = \sigma \underset{\cdot}{\wedge} \gamma$. Consequently, $\sigma \underset{\cdot}{\wedge} \gamma \in \mathrm{id}_\wedge(\Lambda)$.

v. We have $\sigma \underset{\cdot}{\wedge} (\tau \underset{\cdot}{\wedge} \gamma) \leq \sigma \wedge (\tau \underset{\cdot}{\wedge} \gamma) \leq \sigma \wedge \gamma) = (\sigma \wedge \tau) \wedge \gamma$. In particular $\sigma \underset{\cdot}{\wedge} (\tau \underset{\cdot}{\wedge} \gamma) \leq \sigma \wedge \tau$ and $\sigma \underset{\cdot}{\wedge} (\tau \underset{\cdot}{\wedge} \gamma) \leq \gamma$, thus, because $\sigma \underset{\cdot}{\wedge} (\tau \underset{\cdot}{\wedge} \gamma)$ is idempotent (see iv) we obtain $\sigma \underset{\cdot}{\wedge} \gamma) \leq \sigma \underset{\cdot}{\wedge} \tau$ and $\sigma \underset{\cdot}{\wedge} (\tau \underset{\cdot}{\wedge} \gamma) \leq \gamma$, hence, again by iv $\sigma \underset{\cdot}{\wedge} (\tau \underset{\cdot}{\wedge} \gamma) \leq (\sigma \underset{\cdot}{\wedge}) \underset{\cdot}{\wedge} \gamma$. The converse inequality may be derived in formally the same way.

vi. If $x \in \mathrm{id}_\wedge(\Lambda)$ and $x \leq z$, then $x = x \underset{\cdot}{\wedge} z$ follows. A fortiori $x \vee (x \underset{\cdot}{\wedge} \mathrm{z}) = \mathrm{x} = \mathrm{x} \underset{\cdot}{\wedge} z$, and moreover $x \vee (z \underset{\cdot}{\wedge} x) = z \underset{\cdot}{\wedge} x = x$. □

Proposition 2.1
If Λ is a topology of virtual opens, then $\mathrm{id}_{\wedge}(\Lambda)$ *with respect to* $\underset{\cdot}{\wedge}, \vee$, *is a lattice.*

Proof
Direct from the lemma. □

As a consequence we obtain the following, in fact a kind of rephrasing of Proposition 7.1.11 in reference [46].

Corollary 2.1
In the situation of the foregoing proposition and with notation of this section:

i. *If* $x \leq z$ *in* $\mathrm{id}_{\wedge}(\Lambda)$, *then* $x \vee (y \underset{\cdot}{\wedge} z) \leq (z \vee y) \underset{\cdot}{\wedge} z$.

ii. *For* x, y, z *in* $\mathrm{id}_{\wedge}(\Lambda)$ *we obtain* $(x \vee y) \underset{\cdot}{\wedge} (x \vee z) \geq x \vee (y \underset{\cdot}{\wedge} z)$.

The lattice $\mathrm{id}_{\wedge}(\Lambda), \underset{\cdot}{\wedge}, \vee$ *will be denoted by* $SL(\Lambda)$, *and it is called the* ***commutative shadow*** *(lattice) of* Λ. *The term commutative is to the point because it is clear from the definition that* $\sigma \underset{\cdot}{\wedge} \tau = \tau \underset{\cdot}{\wedge} \sigma$.

Observation 2.1
$SL(\Lambda)$ satisfies the modular inequality.

Proof
Take σ, τ, γ in $SL(\Lambda)$ and assume that $\sigma \leq \gamma$. We have to establish that $\sigma \vee (\tau \underset{\cdot}{\wedge} \gamma) \leq (\sigma \vee \tau) \underset{\cdot}{\wedge} \gamma$. The fact that $\sigma \vee (\tau \underset{\cdot}{\wedge} \gamma)$ is idempotent, combined with $\sigma \vee (\tau \underset{\cdot}{\wedge} \gamma) \leq \sigma \vee \tau$ and $\sigma \vee (\tau \underset{\cdot}{\wedge} \gamma) \leq \gamma$, entails $\sigma \vee (\tau \underset{\cdot}{\wedge} \gamma) \leq (\sigma \vee \tau) \wedge \gamma)$, hence $\sigma \vee (\tau \underset{\cdot}{\wedge} \gamma) \leq (\sigma \vee \tau) \underset{\cdot}{\wedge} \gamma$. □

The lattice $SL(\Lambda)$ does not need to be distributive; the noncommutativity of the space Λ prevents this in general; compare this to the fact that the lattice $L(H)$ of closed linear subspaces of a Hilbert space H is not distributive because the projection operators P_U, corresponding to closed linear subspaces U of H, do not commute. On the other hand, the noncommutative topology constructed on Proj$A$ for a schematic graded algebra A (cf. [46][49]), has the lattice of rigid graded torsion theories for its shadow lattice and that is a distributive lattice!

Using intervals as in Exercise 2.1.2 we may define the commutative shadow for an arbitrary noncommutative topology Λ. Define $s : \Lambda \to I(\mathrm{id}_{\wedge}(\Lambda)), \gamma \mapsto \mathrm{id}_{\wedge}(\Lambda) \cap [o, \gamma]$. One easily verifies that $s(\gamma \wedge \sigma) = s(\gamma) \cap s(\sigma) = s(\sigma \wedge \gamma]$. Define $s(\gamma) \vee s(\sigma) = s(\gamma \vee \sigma) = s(\sigma \vee \gamma)$; all claims are easily verified, and we leave this as an exercise because, for our purposes, we will restrict ourselves to the case where $\vee$ is abelian, and then everything is natural and easy.

In Lemma 2.1 we have defined $C(\Lambda)$. When $\wedge$ and $\vee$ are operations making Λ into a noncommutative topology, then we may define $\wedge$ and $\vee$ in $C(\Lambda)$ using the same notation, but that will not lead to confusion, by putting $[A] \wedge [B] = [A \dot{\wedge} B]$, $[A] \vee [B] = [A \dot{\vee} B]$ for $[A], [B] \in C(\Lambda)$, and $A \dot{\wedge} B$ is the directed set $\{a \wedge b, a \in A$ and $b \in B\}$, $A \dot{\vee} B$ is $\{a \vee b, a \in A, b \in B\}$.

Lemma 2.12

If Λ is a noncommutative topology, then $C(\Lambda)$ is a noncommutative topology with respect to the induced operations $\vee, \wedge$, and the canonical poset imorphism $\Lambda \to C(\Lambda)$ is a map of noncommutative topologies.

Proof

Straightforward verification. □

Given a set of $a_i \in \Lambda$, we shall write $p(\wedge, \vee, a_i)$, for a finite bracketed expression involving $\vee, \wedge$ and the letters a_i. For any $\lambda \in \Lambda$ we may look at $W(\lambda)$, the set of all finite length bracketed expressions of type $p(\wedge, \vee, \lambda)$. Obviously, if ω and ω' are in $W(\lambda)$, then $\omega \wedge \omega' \in W(\lambda)$ too. Consequently $[W(\lambda)] \in C(\Lambda)$ is idempotent because $W(\lambda)$ is directed and $W(\lambda)\dot{\wedge}W(\lambda) = W(\lambda)$. Hence when we construct $T(C(\Lambda) = \{p(\wedge, \vee, [A_i]), [A_i] \in i_\wedge(C(\Lambda))\}$ inside $C(\Lambda)$, then this is in some sense "too big" because idempotents like $[W(\lambda)]$ for all $\lambda \in \Lambda$ appear "out of nowhere." In fact, we are mainly interested in stronger idempotents in $C(\Lambda)$, that is, the limits of directed sets containing enough real topological opens. Therefore we say that a directed set A in Λ is **idempotently directed** if for every $a \in A$ there exists an $a' \in A \cap \mathrm{id}_\wedge(\Lambda)$ such that $a' \leq a$. Obviously, when A is idempotently directed in Λ, then [A] is in $\mathrm{id}_\wedge(C(\Lambda))$. We denote by $I_\wedge(C(\Lambda))$ the set $\{[A], A \text{ is idempotently directed}\}$ in Λ. An $[A]$ in $I_\wedge((\Lambda))$ is said to be **strongly idempotent**.

We identify Λ with its image in $C(\Lambda)$ under the canonical poset inclusion $\Lambda \hookrightarrow C(\Lambda); \lambda \mapsto [\{\lambda\}]$. Then observe that $I_\wedge(C(\Lambda)) \cap \Lambda = i_\wedge(\Lambda)$. From the purely abstract point of view it is not necessary to restrict attention to $\mathcal{T}(\Lambda)$; apart from the obvious "constructive" aspect, finite bracketed expressions in idempotent elements of Λ do not enjoy other desirable properties that we are aware of! However, for applications in noncommutative geometry, it is interesting to have open sets covered by finitely many affine ones (these can indeed be defined properly), and that is the only motivation for the consideration of the smaller $\mathcal{T}(\Lambda)$ in Λ. We do obtain an extra problem however: $i_\wedge(C(\mathcal{T}(\Lambda))) \neq i_\wedge(C(\Lambda)$ is possible. On the other hand, we do have that $I_\wedge(C(\mathcal{T}(\Lambda))) = I_\wedge C(\Lambda)$ because $\mathrm{id}_\wedge(T(\Lambda)) = \mathrm{i}_\wedge(\Lambda)$. For any noncommutative topology $C(X)$ we write $\Pi(X)$ for the noncommutative topology obtained by taking the $\wedge$-finite bracketed expressions $P(\wedge, \vee, x_i)$ for strongly idempotent $x_i \in C(X)$. The foregoing observations may thus be summarized in the following.

Lemma 2.13

For a noncommutative topology Λ we have: $\Pi(C(\Lambda)) = \Pi(C(\mathcal{T}(\Lambda)))$.

Proof

Consider $P(\vee, \wedge, [A_i]$ with A_i **idempotently directed** in Λ. By definition of the operations in $C(\Lambda)$, $P(\wedge\vee, [A_i]) = [P(\dot{\wedge}, \dot{\vee}, A_i)]$. The directed set $P(\dot{\wedge}, \dot{\vee}, A_i)$ consists of all bracketed expressions $P(\wedge, \vee, a_i)$ with $a_i \in A_1$; note that if $a_i' \leq a_i$ for all i, then $P(\wedge, \vee, a_i') \leq P(\wedge, \vee, a_i)$; hence $P(\dot{\wedge}, \dot{\vee}, A_i)$ has a cofinal directed subset consisting of all bracketed expressions $P(\vee, \wedge, a_i)$ with $a_i \in A_i \cap i_\wedge(\Lambda)$. Since $P(\vee, \wedge, a_i)$ with $a_i \in A_i \cap i_\wedge(\Lambda)$ are in $T(\Lambda)$ we obtain that $P(\wedge, \vee, [A_i]) \in \Pi$ $(C(T(\Lambda)))$. The desired equality follows easily. □

For notational convenience we shall write τ for $\Pi(C(\Lambda))$; in view of the following proposition τ is again a noncommutative topology; we call it the **pattern topology** or **pattern space** of Λ (or of $C(\Lambda)$). We may think of the pattern space as being defined by a set of $\wedge$-idempotents together with patterns like: $(((-)\wedge(-)\vee(-))\wedge(-), (-)\vee(-), \ldots$. Because certain relations may exist, that is, different patterns evaluated at the same set of elements in $i_\wedge(\Lambda)$ may yield the same result, there is no uniqueness aspect in the description of the pattern topology as described above.

Proposition 2.2
If Λ satisfies the axioms A.1, ..., A.10, hence so does $T(\Lambda)$, then $C(T(\Lambda))$ satisfies the same axioms with respect to the induced operations $\vee, \wedge$ and using $I_\wedge(C(\Lambda))$ instead of $i_\wedge(C(\Lambda))$.

Proof
The axioms A.1, ..., A.8 are trivial to establish because the canonical $T(\lambda) \to C(T(\Lambda))$ respects $\wedge$ and $\vee$.

A.9 Take $[X] \in I_\wedge(C(T(\Lambda)))$ and $[X] \leq [Z]$. If $(x \vee x') \wedge z \in (X\dot{\vee}X)\dot{\wedge}Z$, then we first may select $x_1 \in X$ such that $x_1 \leq x, x_1 \leq x', x_1 \leq z$. Then we choose an idempotent $x_2 \in X$ such that $x_2 \leq x_1$, which is possible because $[X]$ is strongly idempotent. From A.9 in Λ we then obtain $(x_2 \vee x_2) \wedge z \geq x_2 \vee (x_2 \wedge z)$. On the other hand, starting from $x \vee (x' \wedge z)$ in $X\dot{\vee}(X\dot{\wedge}Z)$ we may select an idempotent $x_2 \in X$ such that $x_2 \leq x, x_2 \leq x'$. Hence we obtain $x_2 \vee (x_2 \wedge z) \leq x \vee (x' \wedge z)$, and again from A.9 in Λ it follows that $x_2 \vee (x_2 \wedge z) \leq (x_2 \vee x_2) \wedge z$, the latter representing an element of $(X\dot{\vee}X)\dot{\wedge}Z$.

A.10 Consider a global cover $[X_1] \vee \ldots \vee [X_n] = [\{1\}]$. From $X_1\dot{\vee}\ldots\dot{\vee}X_n \geq \{1\}$ it follows that for any choice of $x_i \in X_i, i = 1, \ldots, n$, we must have $x_1 \vee \ldots, \vee x_n = 1$.

For an arbitrary $[A]$, picking $a \in A$, we obtain $a = (a \wedge x_1) \vee \ldots \vee (a \wedge x_n)$, because of A.10 in Λ. For any $(a_1 \wedge x_1) \vee \ldots \vee (a_n \wedge x_n)$ with $a_i \in A_i, i = 1, \ldots, n$, and $x_j \in X_j, j = 1, \ldots, n$, we may select $a \in A$ such that $a \leq a_i, i = 1, \ldots, n$ and look at $a = (a \wedge x_1) \vee \ldots \vee (a \wedge x_n)$, because of A.10 in Λ. It follows from the foregoing that $[A] = ([A] \wedge [X_1]) \vee \ldots \vee ([A] \wedge [X_n])$. Observe also that $[X] \vee \ldots \vee [X] = [\{1\}]$ yields $x \vee \ldots \vee x = 1$ for every $x \in X$, thus $x = 1$ or $[X] = [\{1\}]$ follows. □

The following observation provides some extra information on global covers and certain operations that may be performed on them.

Observation 2.2
Let Λ be a noncommutative topology.

i. If $1 = \lambda_1 \vee \ldots \vee \lambda_n$, then for every permutation σ of $\{1, \ldots, n\}$, $1 = \lambda_{\sigma(1)} \vee \ldots \vee \lambda_{\sigma(n)}$.

ii. If $1 = \lambda_1 \vee \ldots \vee \lambda_n$, then $1 = \vee_{i,j}\{\lambda_i \wedge \lambda_j, i, j = 1, \ldots, n\}$.

iii. If $\lambda \in i_{\vee}(\Lambda)$ and $\lambda = \lambda_1 \vee \ldots \vee \lambda_n$, write $\lambda_\sigma = \lambda_{\sigma(1)} \vee \ldots \vee \lambda_{\sigma(n)}$, for a permutation σ of $\{1, \ldots, n\}$. Then $\lambda = \lambda_\sigma^{\vee n}$.

In case the cover of λ is induced by a global cover, then $\lambda = \lambda_\sigma$.

Proof

i. Clearly $\lambda_\sigma^{\vee n} \geq \lambda_1 \vee \ldots \vee \lambda_n = 1$, hence $\lambda_\sigma = 1$.

ii. From i we retain that $\lambda_\sigma = 1$. Then A.10 implies that $\lambda_{\sigma(1)} = (\lambda_{\sigma(1)} \wedge \lambda_{\tau(1)}) \vee \ldots \vee (\lambda_{\sigma(1)} \wedge \lambda_{\tau(n)})$, for every permutation τ of $\{1, \ldots, n\}$. Consequently, we obtain $1 = \vee_{i,j}(\lambda_i \wedge \lambda_j)$ (independent of the ordering).

iii. Obviously $\lambda = \lambda^{\vee 2n} \geq \lambda_\sigma^{\vee n} \geq \lambda$, thus $\lambda = \lambda_\sigma^{\vee n}$. In case $\lambda_i = \lambda \wedge \mu_i$ for some global cover $1 = \mu_1 \vee \ldots \vee \mu_n$, then $1 = \mu_{\sigma(1)} \vee \ldots \vee \mu_{\sigma(n)}$ because of i. Consequently, $\lambda = (\lambda \wedge \mu_{\sigma(1)}) \vee \ldots \vee (\lambda \wedge \mu_{\sigma(n)}$ because of A.10 in Λ, thus $\lambda = \lambda_\sigma$. □

Since a noncommutative $\vee$ may be applied to certain questions about electron sets, the general theory of noncommutative topology may be of interest even when in noncommutative geometry, as we deal with it today, the restriction to commutative $\vee$ (or to topologies of virtual opens) is harmless.

In view of Lemma 2.12, $C(\Lambda)$ satisfies in particular A.1, ..., A.9 and so we may consider the commutative shadow SL$(C(\Lambda))$ even in the absence of axiom VOT.3. (anyway this axiom is trivially fulfilled when Λ satisfies FDI). There is a problem in comparing $C(SL(\Lambda))$ and $SL(C(\Lambda))$ because $SL(C(\Lambda))$ is defined on $\mathrm{id}_\wedge(C(\Lambda))$ and every element of $C(SL(\Lambda))$ is obviously coming from an idempotently directed set on $SL(\Lambda)$; moreover, one should carefully note that the operations induced via $(\dot{\wedge})_{.}$ or $(\underset{.}{\wedge})^{\cdot}$ are not a priori comparable. Therefore we define $SL_s(\Pi)$ by restricting first $\underset{.}{\wedge}$ on $i_\wedge(C(\Lambda))$ to $I_\wedge(C(\Lambda))$ and viewing $SL_s(\Pi)$ as the lattice structure induced on $I_\wedge(C(\Lambda))$.

Proposition 2.3
Let Λ be a $\vee$-complete noncommutative topology such that $\vee$ is commutative. Then $SL_s(\Pi) = C(SL(\Lambda))$.

Proof
First observe that if $\vee\mathcal{F}$ exists in Λ for any family $\mathcal{F}$, then $\vee[\mathcal{F}]$ exists in $C(\Lambda)$ for any family $[\mathcal{F}]$ in $C(\Lambda)$. Since $\vee$ is supposed to be commutative, it follows that for every $\mathcal{F}$ in $\mathrm{id}_\vee(\Lambda)$, $\vee\mathcal{F}$ is in $\mathrm{id}_\wedge(\Lambda)$; one can easily see, therefore, that for a family $[\mathcal{F}]$ in $I_\wedge(C(\Lambda))$ again $\vee[\mathcal{F}]$ is in $I_\wedge(C(\Lambda))$. Note that, again from commutativity of $\vee$ or rather because every $\lambda \in \Lambda$ is $\vee$-idempotent, $\vee_f \in \mathcal{F}[A_f] \leq [D]$ in $C(\Lambda)$ is equivalent to $[A_f] \leq [D]$ for all $f \in \mathcal{F}$. For $[A]$, $[B]$ in $I_\wedge(C(\Lambda))$ we want to define $[A]\underset{.}{\wedge}[B]$ as $\vee\{[D], [D] \leq [A] \wedge [B], [D] \in I_\wedge(C(\Lambda))\}$. Now $[D] \leq [A] \wedge [B]$ means $\overline{A \dot{\wedge} B} \subset \overline{D}$; hence for $\wedge$-idempotent $a \in A, b \in B$ there is a $d \in D$, $d \in \mathrm{id}_\wedge(\Lambda)$ such that $d \leq a \wedge b$. So let us define $[A \dot{\wedge} B]$ by the directed system $\{a \underset{.}{\wedge} b, a \in A \cap \mathrm{id}_\wedge(\Lambda), b \in B \cap \mathrm{i}_\wedge(\Lambda)\}$. It is clear by definition that $[A \dot{\wedge} B] \in I_\wedge(C(\Lambda))$

and $[A \dot{\wedge} B] \leq [A]\wedge[B]$. Now, if we define $[A]\dot{\wedge}[B]$ as intended, and observing that the situation $d \leq a \wedge b$ as above carries over to $\bigvee_D d \leq a \wedge b$ with $\bigvee_D d$ still idempotent, then we have obtained that $[A]\dot{\wedge}[B] \leq [A \dot{\wedge} B]$. Therefore we arrive at $[A]\dot{\wedge}[B] = [A \dot{\wedge} B]$ and the claim follows. □

We do not know whether $I_\wedge(\Pi) = I_\wedge(C(\Lambda))$ holds generally. It has been pointed out in [46] that even for the basic example of noncommutative geometry, considering Λ as the noncommutative topology generated by (exact) torsion theories, the question $\mathrm{id}_\wedge(\Lambda) \subset R$-tors is open. Under the stronger restriction that Λ is the noncommutative topology generated by the two-sided Ore sets of a schematic algebra R, the answer is positive, that is, $\mathrm{id}_\wedge(\Lambda)$ in this situation does consist of idempotent torsion theories (but then again maybe not all of these (!), as would be the case for Λ generated by all torsion theories). The good behavior of the Ore set topology seems to be completely due to the technicality that these localizations use elementwise manipulations (not more general left ideal manipulations in the Gabriel topology). We have not yet found a way to extend this type of property to the general axiomatic framework; a possibly far deeper analysis is perhaps contained in the notion of asymmetric topologies (cf. [46], p. 255, $A_{10}^{l,r}$). Obviously, even the ring theory versions of this problem become very abstract torsion theoretic problems; perhaps it is remarkable that the problem vanishes for **schematic algebras**, adding to the unexpected beauty of this class of algebras.

For any poset Λ we say that Λ is **Noetherian** if it satisfies the ascending chain condition; in this generality it is known that Λ is Noetherian if and only if every nonempty subset $\mathcal{Q}$ of Λ has a maximal element. In case Λ is a topology of virtual opens, then Λ is said to be **shadow Noetherian** if the poset $SL(\Lambda)$ is Noetherian, that is, if $\mathrm{id}_\wedge(\Lambda)$ is a Noetherian poset. Obviously if Λ is Noetherian, then it is shadow Noetherian.

Proposition 2.4
With conventions as above:

i. *If Λ is Noetherian, then $C(\Lambda)$ is Noetherian.*

ii. *If Λ is shadow Noetherian, then $I_\wedge(C(\Lambda))$ is Noetherian.*

Proof

i. Consider an ascending chain in $C(\Lambda)$:

(*) $$\cdots \lneqq [A_i] \lneqq [A_{i+1} \lneqq [A_{i+2}] \subset \cdots$$

When passing to corresponding filters we obtain:

(**) $$\cdots \supset \overline{A}_i \supset \overline{A}_{i+1} \supset \overline{A}_{i+2} \supset \cdots$$

From (*) we construct, by selection of elements in A_i, an ascending chain in Λ with $a_i \in \overline{A}_i$

(* * *) $$\cdots \leq a_i \leq a_{i+1} \leq a_{i+2} \leq \cdots$$

There exists a $\gamma_i \in \overline{A}_i$ such that $\gamma_i \leq a_i$ but $\gamma_i \notin \overline{A}_{i+1}$, so we may replace (***) by $\gamma_i \leq a_{i+1} \leq a_{i+2} \leq \ldots$. Again there is a $\gamma_{i+1} \leq a_{i+1}$ such that $\gamma_{i+1} \in \overline{A}_{i+1} - \overline{A}_{i+2}$. Since $\gamma_{i+1} \in \overline{A}_i$ there exists a $\gamma_{i_1} \leq \gamma_i$ with $\gamma_{i_1} \in \overline{A}_i$ and $\gamma_{i_1} \leq \gamma_{i+1}$; then we look at the new chain:

$$\gamma_{i_1} \leq \gamma_{i+1} \leq a_{i+2} \leq \cdots$$

Repetition of the foregoing arguments leads to an ascending chain (the elements being adequately renamed):

$$\gamma_i \leq \gamma_{i+1} \leq \gamma_{i+2} \leq \cdots \leq \gamma_{i+j} \leq \cdots$$

where for all $j = 0, \ldots, n$ we have $\gamma_{i+j} \in \overline{A}_{i+j} - \overline{A}_{i+j+1}$. The Noetherian assumption on Λ implies that the latter chain terminates, hence $\gamma_{i+j} = \gamma_{i+j+1}$ for some finite $j \geq 0$ but that contradicts $\gamma_{i+j} \notin \overline{A}_{i+j+1}$.

ii. If the $[A_i]$ in the chain (*) are in $I_\wedge(C(\Lambda))$, then all choices of elements γ_i in the foregoing proof may be effected in $\text{id}_\wedge(\Lambda)$ and then the shadow-Noetherian property of Λ suffices to arrive at the conclusion. □

Corollary 2.2

If Λ is Noetherian, then the pattern space τ is Noetherian. This follows from i above since any subposet of a Noetherian poset is obviously Noetherian too.

An element of a noncommutative topology, $x \in \Lambda$ say, is $\wedge$**-irreducible** if $x = \lambda_1 \wedge \cdots \wedge \lambda_n$ with $\lambda_i \in \Lambda$ implies $x = \lambda_i$ for some $i \in \{1, \ldots, n\}$. In a similar way we may define $\vee$**-irreducible** elements.

Corollary 2.3

Suppose that Λ is Noetherian.

i. *Every $\lambda \in \Lambda$ is obtained as $\lambda = \rho_1 \wedge \cdots \wedge \rho_n$ with ρ_i $\wedge$-irreducible.*

ii. *Every $\lambda \in \text{id}_\wedge(\Lambda)$ is obtained as $\lambda = \rho_1 \dot{\wedge} \cdots \dot{\wedge} \rho_n$ with $\rho_i \in \text{id}_\wedge(\Lambda)$ being $\dot{\wedge}$-irreducible.*

iii. *For every $[A] \in I_\wedge(C(\Lambda))$ we have that $[A] = [A_1] \dot{\wedge} \cdots \dot{\wedge} [A_n]$ for some $\dot{\wedge}$-irreducible $[A_\alpha] \in I_\wedge(C(\Lambda))$.*

iv. *For every $[A] \in C(\Lambda)$ we have that $[A] = [A_1] \wedge \cdots \wedge [A_n]$ for some $\wedge$-irreducible $[A_\alpha] \in C(\Lambda)$.*

Proof

Since Λ is Noetherian, $C(\Lambda)$, $SL(\Lambda)$ and $I_\wedge(C(\Lambda))$ are also Noetherian posets. The proof of each statement above reduces to a classical argument concerning Noetherian objects; let us phrase it here in case ii. □

Let $\mathcal{F} \subset \text{id}_\wedge(\Lambda)$ be the set of $\lambda \in \text{id}_\wedge(\Lambda)$ that are not a finite $\dot{\wedge}$ of $\dot{\wedge}$-irreducible in $\text{id}_\wedge(\Lambda)$. If $x \in \mathcal{F}$ is a maximal element in $\mathcal{F}$, then $x \in \text{id}_\wedge(\Lambda)$ and it cannot be

$\wedge$-irreducible because otherwise $x = x$ is a $\wedge$ of $\wedge$-irreducibles. Hence $x = x_1 \wedge x_2$ for some $x_1, x_2 \in \mathrm{id}_\wedge(\Lambda)$ with $x \neq x_1, x \neq_2$. From $x < x_1, x < x_2$ it follows that $x_1, x_2 \neq \mathcal{F}$; hence $x_1 = \rho_1 \wedge \ldots \wedge \rho_r$, $x_2 = \rho_{r+1} \wedge \ldots \wedge \rho_n$ for some $\wedge$-irreducible $\rho_i, i = 1 \ldots, n$. Finally $x = \rho_1 \wedge \ldots \wedge \rho_n$ with ρ_i $\wedge$-irreducible in $\mathrm{id}_\wedge(\Lambda)$ follows. □

2.2.2 Projects

2.2.2.1 More Noncommutative Topology

Clearly the definitions and properties contained in Section 2.1 mark only the beginning of a general theory; for purposes here, we do not need to develop this. Starting from a *virtual topology* Λ say, where only the abstract *opens* are given, we may study *closed* sets in $C(\Lambda)$; for example, we define the closure of $[\lambda]$ in $C(\Lambda)$, for $\lambda \in \Lambda$, as follows: $[A] \in [\lambda]^{cl}$, where A is directed in Λ, if $[\mu] \geq [A]$ with $\mu \in \Lambda$ implies that $\mu \wedge \lambda \neq 0$. It is clear that $[\lambda] \in [\lambda]^{cl}$. Similarly one may define the closure of an arbitrary $[B]$ for a directed set B of Λ. This does not make $C(\Lambda)$ into a noncommutative version of a complemented lattice, but it is worthwhile to investigate closed sets and closures.

Compare product pattern topologies and the pattern topology of a product $\Lambda_1 \times \Lambda_2$; compare the commutative shadows of these.

Introduce compactification in $C(\Lambda)$ and generalize relations between compactness and closed sets; study compactness in product topologies as introduced above.

At this point we do not introduce continuous functions (on some sets to be defined). This is postponed until we have introduced quantum topologies, because continuity is expressed on the union of points of an open, that is, viewing an open as a set in some sense.

2.2.2.2 Some Dimension Theory

The dimension theory introduced so far allows application to noncommutative topologies independent of a possible relation to spaces consisting of *points*. In Proposition 2.4 and Corollary 2.2 it was established that the Noetherian assumption on Λ transfers to $C(\Lambda)$ as well as to τ. This allows us to use Krull dimensions (see Lemma 2.6) in each of these cases. In [33] several dimensions have been defined for lattices. These investigations may be carried out for noncommutative, or in particular for virtual topologies, paying particular attention to Krull dimension first. A second part of the project consists of comparing the dimensions, for example, Krull dimensions, of Λ, $C(\Lambda)$, τ in general or in particular cases that will be studied later in these notes.

Lemma 2.14
If $\mathrm{Kdim}\Lambda = 0$, *then* $\mathrm{Kdim}C(\Lambda) = 0$.

Proof
Since Λ has $\mathrm{Kdim}\Lambda = 0$ the interval $[0, 1]$ is Artinian; hence any directed set A has a unique minimal element, say a. If $[A] \leq [B]$, then $\overline{A} \supset \overline{B}$ and $\overline{A} = \{\lambda \in \Lambda, a \leq \lambda\}$, $\overline{B} = \{\mu \in \Lambda, b \leq \mu\}$ where b is the minimal element of B, $a \leq b$. Consequently

$[A] \leq [B]$ if and only if $a \leq b$ and similar with strict inequalities. Any descending chain $\{[A_\alpha], \alpha \in \mathcal{A}\}$ in $C(\Lambda)$ therefore corresponds to a descending chain $\{a_\alpha, \alpha \in \mathcal{A}\}$ in Λ. Since the latter sequence terminates in view of the Artinian property of $[0, 1]$, the original sequence terminates in view of the foregoing remarks, thus $\mathrm{Kdim}C(\Lambda) = 0$. □

In general, if $\mathrm{Kdim}C(\Lambda)$ exists, then we have the inequality $\mathrm{Kdim}(\Lambda) \leq \mathrm{Kdim}C(\Lambda)$.

Lemma 2.15
In case $\mathrm{Kdim}C(\Lambda)$ *exists, then* $\mathrm{Kdim}\Lambda = \mathrm{Kdim}C(\Lambda)$.

Proof
Part of Project B. Hint: if $\mathrm{Kdim}\left[[\overline{A}_{i+1}], [\overline{A}_i]\right] < m$ in $C(\Lambda)$, then for any choice of $a_i \in \overline{A}_i, a_{i+1} \in \overline{A}_{i+1}$ such that $a_{i+1} \leq a_i$, $\mathrm{Kdim}[a_{i+1}, a_i] \leq \mathrm{Kdim}\left[[\overline{A}_{i+1}], [\overline{A}_i]\right]$. The Artinian case may be used as a start of an inductive proof. □

Exercise 2.2

i. *Is* $\mathrm{Kdim}\left[[\overline{A}_{i+1}], [\overline{A}_i]\right]$, *for filters* $\overline{A}_{i+1}, \overline{A}_i$ *such that* $[\overline{A}_{i+1}] \leq [\overline{A}_i]$ *in* $C(\Lambda)$, *obtainable as* $\sup\{\mathrm{Kdim}[a_{i+1}, a_i], a_{i+1} \in \overline{A}_{i+1}, a_i \in \overline{A}_i, a_{i+1} \leq a_i \text{ in } \Lambda\}$?

ii. *If* Λ *has* Kdim, *does it follow in general that* $C(\Lambda)$ *has* Kdim *(then these will be equal)?*

iii. *If* Λ *has* Kdim, *then* $SL(\Lambda)$ *has* Kdim *and the latter is smaller than* $\mathrm{Kdim}\Lambda$; *the difference can be expressed in terms of* $\mathrm{Kdim}[\underset{\cdot}{\lambda}, \lambda]$ *where* $\underset{\cdot}{\lambda}$ *is the largest idempotent smaller than* λ, *for a virtual topology* Λ. *Similarly* $\mathrm{Kdim}\tau(\Lambda)$ *exists and* $\mathrm{Kdim}\tau(\Lambda) \leq \mathrm{Kdim}\Lambda$ *and* $\mathrm{Kdim}\tau(\Lambda) \leq \mathrm{Kdim}C(\Lambda)$.

iv. *Investigate the boundedness condition on a virtual topology* Λ *obtained by demanding that there is an* α *such that* $\mathrm{Kdim}[\underset{\cdot}{\lambda}, \lambda] \leq \alpha$ *for all* $\lambda \in \Lambda$. *This provides a bound on noncommutativity in terms of* Kdim; *verify that* $\mathrm{Kdim}\Lambda \leq \sup\{\mathrm{Kdim}SL(\Lambda), \alpha\}$ *with possible equality when* $\alpha = \sup_{\lambda \in \Lambda} \mathrm{Kdim}[\underset{\cdot}{\lambda}, \lambda]$.

2.3 Points and the Point Spectrum: *Points in a Pointless World*

We say that $[A]$ in $C(\Lambda)$ is a **quasi-point** of $C(\Lambda)$ (or of Λ) if $\overline{A}$ is a maximal filter, that is $\overline{A} \neq \Lambda$ but if $\overline{A} \subsetneq B$ where B is a filter, then $B = \Lambda$.

Lemma 2.16
If $[A]$ *is a quasi-point in* $C(\Lambda)$, *then it is in* $\mathrm{id}_\wedge(C(\Lambda))$ *and it is a minimal nonzero element in the poset* $C(\Lambda)$.

Proof

If $[A] \wedge [A] \neq [A]$, then the filter defining $[A] \wedge [A]$ is strictly containing $\overline{A}$; hence it must be the whole of Λ, meaning $[A] \wedge [A] = 0$ or $[A] = 0$. □

Since quasi-point is a confusing term we prefer to refer to them as **minimal-points**, or later also as **pointed filters**.

The definition of **points** we intend to give is inspired by the way prime ideals of commutative rings appear as elements of the prime spectrum in scheme theory. In fact it will turn out that there is a nice noncommutative equivalent for this in noncommutative geometry when phrased in terms of prime torsion theories.

Lemma 2.17

The following statements are equivalent:

i. *In $C(\Lambda)$, $[A] \leq \vee\{[A_\alpha], \alpha \in \mathcal{A}\}$ entails $[A] \leq [A_\alpha]$ for some $\alpha \in \mathcal{A}$.*

ii. *If $\vee\{\lambda_\alpha, \alpha \in \mathcal{A}\} \in \overline{A}$ in Λ, then $\lambda_\alpha \in \overline{A}$ for some $\alpha \in \mathcal{A}$.*

Proof

Let D be the filter in Λ defined by the $\vee\{\lambda_\alpha, \alpha \in \mathcal{A}\}$ for varying $\lambda_\alpha \in A_\alpha$. Then $d \in D$ if $d \geq \vee\{a_\alpha, \alpha \in \mathcal{A}\}$ for some $a_\alpha \in A_\alpha$; hence $d \geq a_\alpha$ for all $\alpha \in \mathcal{A}$ or $d \in \overline{A}_a$ for all $\alpha \in \mathcal{A}$. Thus $D \subset \cap_\alpha\{\overline{A}_\alpha, \alpha \in \mathcal{A}\}$. Conversely, if $x \in \cap_\alpha\{\overline{A}_\alpha, \alpha \in \mathcal{A}\}$, then $x \in \overline{A}_\alpha$ for all $\alpha \in \mathcal{A}$ and hence $x = x \vee \ldots \vee x \in D$. Therefore $D = \cap_\alpha\{\overline{A}_\alpha, \alpha \in \mathcal{A}\}$ and $[D] = \vee\{[A_\alpha], \alpha \in \mathcal{A}$ follows.

i. $\Rightarrow$ ii. Suppose $\vee\{\lambda_\alpha, \alpha \in \mathcal{A}\} \in \overline{A}$ for some $\lambda_\alpha \in \Lambda$ and define filters $\overline{A}_\alpha = \{b \in \Lambda, b \geq \lambda_\alpha\}$. Then obviously, $\cap\{\overline{A}_\alpha, \alpha \in \mathcal{A}\} = \{b \geq \vee\{\lambda_\alpha, \alpha \in \mathcal{A}\}\}$. By the introductory remarks i reduces to the condition $\cap\{\overline{A}_\alpha . \alpha \in \mathcal{A}\} \subset \overline{A}$ implies $\overline{A}_\alpha \subset \overline{A}$ for some $\alpha \in \mathcal{A}$. Now $\vee\{\lambda_\alpha, \alpha \in \mathcal{A}\} \in \overline{A}$ entails $\cap\{\overline{A}_\alpha, \alpha \in \mathcal{A}\} \subset \overline{A}$, hence $\overline{A}_\alpha \subset A$ or $\lambda_\alpha \in \overline{A}$ follows for some $\alpha \in \mathcal{A}$.

ii. $\Rightarrow$ i. From $\cap\{\overline{A}_\alpha, \alpha \in \mathcal{A}\} \subset \overline{A}$ it follows that for all choices of $\lambda_\alpha \in \overline{A}_\alpha$ we have $\vee\{\lambda_\alpha, \alpha \in \mathcal{A}\} \in \overline{A}$. Assumption ii then yields that $\lambda_\alpha \in \overline{A}$ for some $\alpha \in \mathcal{A}$. If $\overline{A}_\alpha \not\subset \overline{A}$ for all $\alpha \in \mathcal{A}$, choose $a_\alpha \in \overline{A} - \overline{A}_\alpha$ and look at $\vee\{a_\alpha, \alpha \in \mathcal{A}\} \in \overline{A}$; then ii leads to a contradiction. □

Definition 2.2

If $[A]$ satisfies one of the equivalent conditions of the foregoing lemma, then we say that $[A]$ is **a point** of Λ (or $C(\Lambda)$. This is unambiguous because we do not consider points in $C(C(\Lambda)))$. We define the **point-spectrum** $Sp(\Lambda) = \{[p], [p]$ is a point of $\Lambda\}$. Put $p(\lambda)$ for λ in Λ equal to $\{[p] \in Sp(\Lambda), [p] \leq [\lambda\}$, then $p(\lambda \wedge \mu) \subset p(\lambda) \cap p(\mu)$, $p(\lambda \vee \mu) = p(\lambda) \cup p(\mu)$. Indeed, for the latter observe that $[p] \leq [\lambda \vee \mu] = [\lambda] \vee [\mu]$ implies $[p] \leq [\lambda]$ or $[p] \leq [\mu]$. Therefore, putting $U_\lambda = p(\lambda)$ in $Sp(\Lambda)$ we have obtained a basis for a topology on $Sp(\Lambda)$, called the **point-topology**. Write $SP(\Lambda)$ for $Sp(\Lambda) \cap I_\wedge(C(\Lambda))$; we refer to this as the **Point-spectrum** (with capital P). For λ in Λ we consider $P(\lambda)$ in $C(\Lambda)$ given by $\{[P] \in SP(\Lambda), [P] \leq [\lambda]\} = p(\lambda) \cap I_\wedge(C(\Lambda))$. The topology induced on $SP(\Lambda)$

by the point topology of $Sp(\Lambda)$ is referred to as the **Point-topology**. Observe that $P(\lambda \wedge \mu) = P(\lambda) \cap P(\mu)$ because if an indempotent $[p]$ is such that $[p] \leq [\lambda]$ and $[p] \leq [\mu]$, then also $[p] \leq [\lambda] \wedge [\mu] = [\lambda \wedge \mu]$; therefore the opens $U_\lambda \cap I_\wedge(C(\Lambda))$ are exactly the $P(\lambda)$ and the latter form a topology on $SP(SL(C(\Lambda)))$.

Let us come back to the FDI property and contracting idempotents (see remarks after Section 2.2.1 (Properties). A noncommutative topology Λ is said to satisfy the **weak *FDI* property** if every $\lambda \in \mathrm{id}_\wedge(\Lambda)$ is contracting.

Lemma 2.18
If Λ satisfies the weak FDI property, then every $[A] \in I_\wedge(C(\Lambda))$ is contracting.

Proof
Take $[A] \in I_\wedge(C(\Lambda))$ and suppose that $[A] \leq [B] \vee [C]$. The claim will follow if we establish:

$$(*) \qquad [A] \leq ([A] \wedge [B]) \vee ([A] \wedge [C])$$

(the proof with respect to $\bigvee_{\alpha \in \mathcal{A}}$ is similar).

Take $(a \wedge b) \vee (a' \wedge c)$ in the directed set defining the righthand side of (*). Since $[A] \leq [B] \cup [C]$ there exists an $x \in A$ such that $x \leq b \vee c$, and since $[A] \in I_\wedge(C(\Lambda))$ there exists an idempotent $a'' \in \mathrm{id}_\wedge(\Lambda)$ such that $a'' \in \overline{A}$, $a'' \leq a$, $a'' \leq a'$, $a'' \leq x$. Now from $a'' \leq x$ with $a'' \in \mathrm{id}_\wedge(\Lambda)$ it follows that $a'' = (a'' \wedge b) \vee (a'' \wedge c)$ in view of the assumed weak FDI property. Then from the inequality:

$$a'' = (a'' \wedge b) \vee (a'' \wedge c) \leq (a \wedge b) \vee (a' \wedge c)$$

it follows that $(a \wedge b) \vee (a' \wedge c) \in \overline{A}$. Hence the directed set, and in fact also the filter corresponding to $([A] \wedge [B]) \vee ([A] \wedge [C])$ is contained in $\overline{A}$; then we arrive at (*). □

Proposition 2.5
Suppose Λ satisfies the weak FDI property; then $[A] \in I_\wedge(C(\Lambda))$ is a Point if and only if it is $\vee$-irreducible.

Proof
If $[A] \leq \{[A_\alpha], \alpha \in \mathcal{A}\}$, then the foregoing lemma yields $[A] = \vee\{[A] \wedge [A_\alpha], \alpha \in \mathcal{A}\}$ and the statement is clear. □

Corollary 2.4
For Λ as in the proposition, if $[F] \in I_\wedge(C(\Lambda))$ is a quasi-point, then it is a Point. Indeed a quasi-point being minimal nonzero is automatically $\vee$-irreducible.

Corollary 2.5
If Λ satisfies the weak FDI property and is $\vee$-complete, then the pattern topology is a topology of virtual opens (as always, we restricted to commutative $\vee$).

We define $Sp(\tau)$ and $SP(\tau) = Sp(\tau) \cap I_{\wedge}(C(\Lambda))$ and endow them with the topology induced by the point-topology of Λ to obtain, respectively, the **point-pattern topology** and the **Point-pattern topology**; we shall refer to these sets as the **pattern spectrum**, respectively, the **pattern-Spectrum.**

To finish this section let us point out that minimal points are points under different conditions sometimes. We say that Λ is **converging-distributive** if for $\lambda_1, \ldots, \lambda_n \in \Lambda$ there exist nonzero $\lambda'_1 \leq \lambda_1, \ldots, \lambda'_n \leq \lambda_n$ such that $(\underset{\alpha}{\vee} \lambda'_\alpha) \wedge \mu \subset \underset{\alpha}{\vee}(\lambda_\alpha \wedge \mu)$.

Proposition 2.6
Suppose that Λ is converging-distributive.

i. Any minimal-point is a point.

ii. Any $[p] \in C(\Lambda)$ is a point if and only if it is a $\vee$-irreducible element in $C(\Lambda)$.

Proof

i. If $[A]$ represents a minimal-point, then $\overline{A}$ is a maximal filter ($\neq \Lambda$). Assume that $\cap\{\overline{A}_\alpha, \alpha \in \mathcal{A}\} \subset \overline{A}$ and $\overline{A}_\alpha \not\subset \overline{A}$ for all α. Pick $a_\alpha \in \overline{A}_\alpha - \overline{A}$ for every $\alpha \in \mathcal{A}$. Look at $B_\alpha = \{a_\alpha \wedge a, a \in \overline{A}\}$; this is obviously a directed set in Λ. Moreover $\overline{B}_\alpha \supset \overline{A}$ because for every $a \in \overline{A}$ there is $a_\alpha \wedge a \in B_\alpha$ such that $a_\alpha \wedge a \leq a$. Since $a_\alpha \in \overline{B}_\alpha - \overline{A}$, it follows that $\overline{B}_\alpha = \Lambda$; hence $a_\alpha \wedge a = 0$ for some $a \in \overline{A}$. In view of the converging-distributivity of Λ we may select $a'_\alpha \leq a_\alpha$ such that $(\vee_{\alpha \in \mathcal{A}} a'_\alpha) \wedge a \subset \vee_{\alpha \in \mathcal{A}}(a_\alpha \wedge a)$. Here $a'_\alpha \leq a_\alpha$ implies $a'_\alpha \notin \overline{A}$ and $\vee_{\alpha \in \mathcal{A}} a'_\alpha \in \cap_{\alpha \in \mathcal{A}} \overline{A}_\alpha \subset \overline{A}$. Thus we obtain $(\vee_{\alpha \in \mathcal{A}} a'_\alpha) \wedge a = 0$, but the latter is in $\overline{A}$ or $o \in \overline{A}$ contradicting the assumptions on $\overline{A}$.

ii. If $[p] \leq \vee\{[A_\alpha], \alpha \in \mathcal{A}\}$, then $[p] = \vee\{[p] \wedge [A_\alpha], \alpha \in \mathcal{A}\}$ follows from converging distributivity in a straightforward way. □

Since neither points nor minimal points need to exist in a given noncommutative or virtual topology, we may look for another useful generalization of the notion of point. The generalized points should then in some way allow us to describe the noncommutative topology; that is, there should be enough generalized points so that an open can be somewhat characterized by a set of generalized points. However, we certainly do not want to return to a set theoretical definition of the topology! By now it should be clear to the reader that the noncommutativity of space is essentially linked to its non-set theoretical nature! Nevertheless, we are inspired by the desire to define a consistent concept of quantum space carrying a quantum topology that could be a tool in the physics of a noncommutative world (see Section 4.3). So we look for a notion of generalized point allowing us to describe the opens up to some undetermined point stemming from noncommutativity, that is, up to iterated self-intersections! The generalization should at least have the effect that a space having enough of these generalized points benefits from this in the sheaf theory, that is, in the construction of a sheafification technique.

Consider a $\vee$-complete topology Λ; in fact it is harmless to restrict ourselves to virtual topologies here. A **partition** for Λ is a family $\{b_\alpha \neq 0, \alpha \in \mathcal{A}\}$ in Λ such that $\vee\{b_\alpha, \alpha \in \mathcal{A}\} = 1$ and $b_\alpha \wedge b_\beta = 0$ for every $\beta \neq \alpha$ in $\mathcal{A}$. We say this such that a partition is a **generating partition for** Λ if for all $\lambda \in \Lambda$ there are $n_\alpha \in \mathbb{N}$ for $\alpha \in \mathcal{B}$, some $\mathcal{B}$ a subset of $\mathcal{A}$, such that $\lambda = \vee\{b_\alpha^{n_\alpha}, \alpha \in \mathcal{B}\}$, where b_α^m will denote the m-fold self-intersection of b_α, that is, $b_\alpha^m = b_\alpha \wedge \ldots \wedge b_\alpha$ with m-terms appearing in the righthand side of the expression. On the set of partitions for λ we define a relation $<$ by stating that $\{b_\alpha, \alpha \in \mathcal{A}\} < \{c_\gamma, \gamma \in \Gamma\}$ if there exists a map $m : \Gamma \to \mathcal{A}$ such that $c_\gamma \leq b_{m(\gamma)}$ for $\gamma \in \Gamma$; then we say that $\{c_\gamma, \gamma \in \Gamma\}$ is a **refinement** of $\{b_\alpha, \alpha \in \mathcal{A}\}$.

Proposition 2.7
Let Λ be a $\vee$-complete noncommutative topology with commutative $\vee$.

i. *If Λ has a generating partition, then it is unique.*
ii. *If a generating partition has a refinement that contains a finite global cover, then the generating partition equals that refinement and it is a finite set.*
iii. *If Λ satisfies axiom A.10 also with respect to possibly infinite global covers, then a generating partition does not allow a proper refinement in the set of partitions; however, in this case the existence of a generating partition makes Λ commutative (every $\lambda \in \Lambda$ is $\wedge$-idempotent).*

Proof

i. Consider generating partitions $\{b_\alpha, \alpha \in \mathcal{A}\}$ and $\{c_\gamma, \gamma \in \Gamma\}$. Then for $\alpha \in \mathcal{A}$, $b_\alpha = \vee\{c_\gamma^{n_{\gamma\alpha}}, \gamma \in \mathcal{B}_\alpha\}$ for some $\mathcal{B}_\alpha \subset \Gamma$ and $n_{\gamma\alpha} \in \mathbb{N}$. On the other hand we also have that
$$c_\gamma^{n_{\gamma\alpha}} = \vee\{b_\delta^{n_{\delta\gamma}}, \delta \in \mathcal{B}_\gamma\} \text{ for some } \mathcal{B}_\gamma \subset \mathcal{A}, n_{\delta\gamma} \in \mathbb{N}.$$
From $b_\delta^{n_{\delta\gamma}} \leq c_\gamma^{n_{\gamma\alpha}} \leq b_\alpha$ it follows that $\delta = \alpha$; hence we arrive at:
$$c_\gamma^{n_{\gamma\alpha}} = b_\alpha^{n_{\delta\gamma}}, \quad \gamma \in \mathcal{B}_\alpha \tag{$*$}$$
Using the generating property for $\{b_\alpha, \alpha \in \mathcal{A}\}$ we can write $c_\gamma = \vee\{b_\delta^{n_\delta}, \delta \in \mathcal{B}'_\gamma\}$ for some $\mathcal{B}'_\gamma \subset \mathcal{A}$, $n_\delta \in \mathbb{N}$. If $\delta \in \mathcal{B}'_\gamma$ then we have:
$$b_\delta^{n_\delta} \leq c_\gamma; \text{ hence } b_\delta^d \leq c_\gamma^{n_{\gamma\alpha}} \text{ for } d = n_\delta n_{\gamma\alpha}.$$
Combined with (*) this leads to $\delta = \alpha$ and $c_\gamma = b_\alpha^{n_\alpha}$, thus $c_\gamma \leq b_\alpha$. Now (*) holds for all $\gamma \in B_\alpha$ but since the $\wedge$-powers of b_α cannot be disjoint there can be only one $\gamma \in B_\alpha$ and $b_\alpha = c_\gamma^{n_{\gamma\alpha}}$, or $b_\alpha \leq c_\gamma$. Finally $b_\alpha = c_\gamma$ and the correspondence $\alpha \leftrightarrow \gamma$ is clearly a bijection $\mathcal{A} \leftrightarrow \Gamma$.

ii. Consider a refinement $\{c_\gamma, \gamma \in \Gamma\}$ of the generating partition $\{b_\alpha, \alpha \in \mathcal{A}\}$ given by the map $m : \Gamma \to \mathcal{A}$. For $\gamma \in \Gamma$ with $m(\gamma) = \alpha$ we have $c_\gamma \leq b_\alpha$.

From $c_\gamma = \vee\{b_\alpha^{n_\alpha}, \alpha \in \mathcal{B}\}$ for some $\mathcal{B} \subset \mathcal{A}$ and $n_\alpha \in \mathbb{N}, \alpha \in \mathcal{B}$ we see that $\beta \neq \alpha \in \mathcal{B}$ would lead to $b_\beta^{n_\beta} \leq c_\gamma \leq b_\alpha$ or $b_\beta^{n_\beta} = 0$; hence $b_\beta = 0$, a contradiction. Therefore $c_\gamma = b_\alpha^{n_\alpha}$ and $m(\gamma) = \alpha$ for a unique γ. Use the bijection $\Gamma \to \mathrm{Im}(m)$ to reindex the set $\{c_\gamma, \gamma \in \Gamma\}$ such that $c_\alpha \leq b_\alpha$; that is, we use m as an identification of Γ and $\mathrm{Im}(m)$. If $\{c_\gamma, \gamma \in \Gamma\}$ contains a finite global cover, say $1 = c_{\gamma_1} \vee \ldots \vee c_{\gamma_d}$, then $1 = b_{\gamma_1}^{n_1} \vee \ldots \vee b_{\gamma_d}^{n_d}$ for some $n_1, \ldots, n_d \in \mathbb{N}$. Applying axiom A.10 to b_α with $\alpha \neq \gamma_1, \ldots, \gamma_d$ yields $b_\alpha = (b_\alpha \wedge b_{\gamma_1}^{n_1}) \vee \ldots \vee (b_\alpha \wedge b_{\gamma_d}^{n_d})$, hence $b_\alpha = 0$, a contradiction. Consequently $\mathcal{A} = \{\gamma_1, \ldots, \gamma_d\}$. Moreover $b_{\gamma_i} = \vee_j (b_{\gamma_i} \wedge b_{\gamma_j}^{n_j}) = b_{\gamma_i} \wedge b_{\gamma_i}^{n_i} \leq b^{n_i} = c_{\gamma_i}$. Together with $c_{\gamma_i} = b_{\gamma_i}^{n_i}$ derived above, that is, $c_{\gamma_i} \leq b_{\gamma_i}$ this yields $c_{\gamma_i} = b_{\gamma_i}$ proving that $\{c_\gamma, \gamma \in \tau\}$ and $\{b_\alpha, \alpha \in \mathcal{A}\}$ are the same (and finite).

iii. Let $\{c_\gamma, \gamma \in \Gamma\}$ be a refinement of $\{b_\alpha, \alpha \in \mathcal{A}\}$ via the map $m : \Gamma \to \mathcal{A}$. As before, we arrive at $c_\alpha = b_\alpha^{n_\alpha}$ for some $n_\alpha \in \mathbb{N}$ (up to identifying Γ and $m(\Gamma)$ in $\mathcal{A}$). Applying axiom A.10 in its infinite form, the global cover $1 = \vee\{b_\alpha^{n_\alpha}, \alpha \in \Gamma\}$ induces for any $b_\tau, \tau \in \mathcal{A}$, $b_\tau = \vee\{b_\tau \wedge b_\alpha^{n_\alpha}, \alpha \in \Gamma\}$. First it is clear that m has to be surjective because if $\tau \notin \mathrm{Im}(m)$, then $b_\tau = 0$, a contradiction. Furthermore $b_\tau \wedge b_\alpha^{n_\alpha} = 0$ unless $\alpha = \tau$, hence $b_\tau = b_\tau^{n_\tau + 1}$, thus $b_\tau \leq c_\tau \leq b_\tau$ leads to $c_\tau = b_\tau$ for all $\tau \in \mathcal{A}$. We establish that $\mathcal{A}$ is bijective to Γ (via m) and $\{c_\gamma, \gamma \in \Gamma\} = \{b_\alpha, \alpha \in \mathcal{A}\}$. However we get more, $b_\alpha = \vee_\beta\{(b_\alpha \wedge b_\beta), \beta \in A\}$ yields $b_\alpha = b_\alpha \wedge b_\alpha$ or $b_\alpha \in \mathrm{id}_\wedge(\Lambda)$ for any $\alpha \in \mathcal{A}$. For $\lambda \neq 0$ in Λ we thus find $\lambda = \vee\{b_\alpha, \alpha \in \mathcal{B}\}$ for some $\mathcal{B} \subset \mathcal{A}$ and then we obtain $\lambda \wedge \lambda \geq b_\alpha \wedge b_\alpha = b_\alpha$ for all $\alpha \in \mathcal{B}$. Commutativity of $\vee$ yields $\lambda \wedge \lambda \geq \vee\{b_\alpha, \alpha \in \mathcal{B}\} = \lambda$ or $\lambda = \lambda \wedge \lambda$ and $\lambda \in \mathrm{id}_\wedge(\Lambda)$. □

One should not be demotivated by the last statement of Proposition 2.7 (iii). In fact we do not want "points" to be elements of Λ but rather of $\mathcal{C}(\Lambda)$. Also observe that Proposition 2.7(i) is independent of the axiom A.10.

Definition 2.3

We say that a noncommutative topology Λ with commutative $\vee$ (and being $\vee$-complete) has a **weak quantum basis** $\{[B_\alpha], \alpha \in \mathcal{A}\}$ if this is a partition for $C(\Lambda)$ that is Λ**-generating** in the sense that for each $\lambda \in \Lambda$ we have $[\lambda] = \vee\{[B_\alpha]^{n_{\alpha\lambda}}, \alpha \in \mathcal{B}_\lambda\}$ for $\mathcal{B}_\lambda \subset \mathcal{A}$ and $n_{\alpha\wedge} \in \mathbb{N}$. We say that Λ has a **quantum basis** if $\mathcal{C}(\Lambda)$ has a generating partition; in that case such a quantum basis is unique and its elements $[B_\alpha]$ are called **quantum points**.

Example 2.2

In the fundamental example treated in Section 2.7., that is, the lattice $L(H)$ for a Hilbert space H, the quantum points are exactly the minimal points.

Proposition 2.8

Let Λ be as in Proposition 2.3.8 and suppose a quantum basis $\{[B_\alpha], \alpha \in \mathcal{A}\}$ exists. All minimal points of Λ are exactly the $\wedge$-idempotent $[B_\alpha]^{n_\alpha}$ corresponding to those $[B_\alpha]$ for which the interval $[[0], [B_\alpha]]$ in $C(\Lambda)$ is finite. The $[B_\alpha]$ such that no $[B_\alpha]^n$ is a minimal point define an interval $[[0], [B_\alpha]]$ containing exactly the different $[B_\alpha]^n$.

Proof
We have observed that any $[A] \leq [B_\alpha]$ is necessarily of the form $[B_\alpha]^{n_\alpha}$ for some $n_\alpha \in \mathbb{N}$. If $[A] \leq [B_\alpha]$ is a nonzero $\wedge$-idempotent in $C(\Lambda)$, then $[A] = [B_\alpha]^{n_\alpha}$ and $[B]_\alpha^{n_\alpha} = [B]_\alpha^{2n_\alpha}$; thus $[B_\alpha]^{n_\alpha} \geq [B_\alpha]^{n_\alpha+1} \geq [B_\alpha]^{2n_\alpha}$ implies that $[B_\alpha]^{n_\alpha+d} = [B_\alpha]^{n_\alpha}$ for every $d \in \mathbb{N}$; thus $[[0], [B_\alpha]]$ is finite and $[B_\alpha]^{u_\alpha}$ is a minimal element different from $[0]$ in $C(\Lambda)$, thus a minimal point for Λ. Conversely, if $[[0], [B_\alpha]]$ is finite, then for some $n_\alpha \in \mathbb{N}$, $[B_\alpha^{n_\alpha}] \neq [0]$ is a minimal element of $C(\Lambda)$; therefore it is in particular $\wedge$-idempotent. In case $[B_\alpha]$ is such that $[B_\alpha]^n = [B_\alpha]^m$ for some $n > m$ in $\mathbb{N}$, then there exists an $n_\alpha \in \mathbb{N}$ such that all $[B_\alpha]^d$ with $d \leq n_\alpha$ are different in $C(\Lambda)$ but $[B_\alpha]^{n_\alpha} = [B_\alpha]^{n_\alpha+k}$ for any $k \in \mathbb{N}$. Therefore we are either in the situation where all $[B_\alpha]^d$ for $d \in \mathbb{N}$ are different constituting the interval $[[0], [B_\alpha]]$ and none of the $[B_\alpha]^d$ is $\wedge$-idempotent, or else $[[0], [B_\alpha]]$ is finite and $[B_\alpha]^{n_\alpha}$ is $\wedge$-idempotent (and a minimal point) for some $n_\alpha \in \mathbb{N}$. □

To every $\lambda \in \Lambda$ as in Proposition 2.8, there corresponds a set of quantum points $\{[B_\alpha], \alpha \in \mathcal{A}_\lambda\}$ obtained from the decomposition $[\lambda] = \vee_{\alpha \in \mathcal{A}_\lambda} \{[B_\alpha]^{n_\alpha}, \alpha \in \mathcal{A}_\lambda\}$ some $\mathcal{A}_\lambda \subset \mathcal{A}$; we denote that set by $S(\lambda)$. The set $S(\lambda)$ does not define λ; it defines uniquely the quantum closure $\lambda^{\text{cl}} = \vee \{[B_\alpha], \alpha \in \mathcal{A}_\lambda\}$, but $\lambda \leq \lambda^{\text{cl}}$ cannot be reconstructed unless we know the "multiplicities" n_α, which measure in some sense the noncommutativity of Λ; also these numbers measure how quantum points fail to be $\wedge$-idempotent. A $[B_\alpha]$ not dominating any nonzero $\wedge$-idempotent, that is, if $[A] \neq 0$ is such that $[A] \leq [B_\alpha]$, then $[A]$ is not idempotent in $C(\Lambda)$, which is called a **radical point** (short for radical quantum point).

More generally in any Λ an element $\lambda \neq 0$ such that λ does not dominate a nonzero idempotent is called a **radical element**. If $[A]$ is a radical element of $C(\Lambda)$, then $[A] = \vee \{[B_\alpha]^{n_\alpha}, \alpha \in \mathcal{A}_A\}$ for some $n_\alpha \in \mathbb{N}$, where all $[B_\alpha], \alpha \in \mathcal{A}_A$ are radical points.

2.3.1 Projects

2.3.1.1 The Relation between Quantum Points and Strong Idempotents

The relation between quantum points and strong idempotents in $I_\wedge(C(\Lambda))$ is enigmatic to say the least; therefore mixing conditions expressing generation properties in terms of strong idempotents (using bracketed expressions in $\wedge$ and $\vee$) and in terms of quantum points (using only expressions in $\vee$) must lead to complex situations. The final definition in this philosophy should be that of a **quantum topology**: a topology of virtual opens such that τ, the pattern topology, has a quantum basis (the elements of this should then be logically termed **quantum patterns**; what's in a name?). The intrinsic technical problem is that we do not have, in the generality we consider here, any prescribed relation between the intervals $[[0], [A]]$ and $[[B], [A] \vee [B]]$ in $C(\Lambda)$. One could put conditions in terms of Kdim, composition series involving critical elements, composition series involving $\wedge$-powers of quantum points, etc. The topologies in classical examples are defined on sets with a lot of extra geometric or algebraic structure (e.g., groups, vector spaces); without fixing such extra structure one should try to find suitable abstract "lattice theory–type" properties that can replace the missing structure.

2.3.1.2 Functions on Sets of Quantum Points

Generalized points suggest a beginning of spectrum theory, implications in sheaf theory (we shall come back to this later), and the use of set theoretic functions. If $\mathcal{A}$ is the set of quantum points (as before $[B_\alpha]$ with $\alpha \in \mathcal{A}$), then $C(\Lambda)$ may be obtained from functions $n : \mathcal{A} \to \mathbb{N}, \alpha \mapsto n_\alpha$, corresponding to n the element $\vee\{[B_\alpha]^{n_\alpha}, \alpha \in \mathcal{A}$ such that $n(\alpha) \neq 0\}$. Putting $[B_\alpha]^o = [0]$ we obtain a map $\mathcal{F}(\mathcal{A}, \mathbb{N}) \to C(\Lambda)$, which is surjective. We may also study maps between noncommutative topologies Λ and Λ' as before, defined by functions on the sets of quantum points. Such functions do not necessarily lead to a well-defined map from $C(\Lambda)$ to $C(\Lambda')$ because the expression of $[A] \in C(\Lambda)$ as a $\vee$ of $\wedge$-powers of quantum points may not be unique; some care is necessary here. Determining formal links between sets of quantum points and functions on these and noncommutative topological features is an interesting project. Obvious links to sheaf theory and spectral theory can be exploited. Using functions to sets with extra structures may lead to interesting examples carrying algebra-like structures; for example, look at the appearance of $\mathbb{P}(H)$, the projective Hilbert space, in the case of $L(H)$ studied in Section 2.7, as an effect of the linearity of the theory.

2.4 Presheaves and Sheaves over Noncommutative Topologies

In this section we consider a noncommutative topology Λ. In view of Proposition 2.2 and Lemma 2.12, $T(\Lambda)$, $C(\Lambda)$, τ are noncommutative topologies too.

We are given a category $\underline{C}$ and we are interested in sheaves over Λ, $C(\Lambda)$, $\tau, \ldots$ and relations among these; the category $\underline{C}$ is rather arbitrary. $\underline{C}$ must have sums, products, limits, and its objects are assumed to be sets. In Section 2.1.3, separated presheaves and sheaves over general posets were introduced. We apply these to Λ, $C(\Lambda)$, $T(\Lambda)$, $\tau, \ldots$

Given a presheaf $P : \underline{\Lambda}^o \longrightarrow \underline{C}$. Consider $[A] \in C(\Lambda)$ and define: $P_{[A]} = \varinjlim_{a \in A} P(a)$ in $\underline{C}$. Observe that for $B \sim A$ we do obtain that $\varinjlim_{a \in A} P(a) = \varinjlim_{b \in B} P(b)$, so that the definition of $P_{[A]}$ is well defined. So we may look at $P_{[\quad]} : C(\Lambda)^o \to \underline{C}$ defined by $P_{[\quad]}([A]) = P_{[A]}$ for all $[A] \in C(\Lambda)$.

Lemma 2.19

If P is a presheaf, then $P_{[\quad]}$ is a presheaf on $C(\Lambda)$.

Proof

If $[A] \leq [B]$, then for all $b \in B$ there is an $a \in A$ such that $a \leq b$; hence there is a canonical map $\varinjlim_{b \in B} P_b \to \varinjlim_{a \in A} P_a$, induced by the restriction maps $\rho_a^b : P_b \to P_a$ defined by the presheaf P. We shall write $\rho_{[A]}^{[B]}$ for the map defined before. In case we have $[B] = [A]$ we do obtain that $\rho_{[A]}^{[B]}$ is the identity map $1_{P_{[A]}}$. For $[C] \leq [B] \leq [A]$ it is easy enough to check that the usual composition formula holds: $\rho_{[C]}^{[B]} \rho_{[B]}^{[A]} = \rho_{[C]}^{[A]}$.

□

Corollary 2.6
Let $P_{[\ ]}|\Lambda$ be the restriction of $P_{[\ ]}$ to Λ via the canonical inclusion $\Lambda \hookrightarrow C(\Lambda)$, $\lambda \mapsto [\lambda]$, then $P_{[\ ]}|\Lambda = P$.

Theorem 2.1
With notation and conventions as before.

i. *If P is a separated presheaf, then $P_{[\ ]}$ is a separated presheaf.*

ii. *If P is a sheaf over Λ, then $P_{[\ ]}$ is a sheaf over $C(\Lambda)$.*

Proof

i. Let P be a separated presheaf over Λ and look at a finite cover $\{[A_\alpha], \alpha \in \mathcal{A}\}$ of $[A]$ in $C(\Lambda)$. Suppose that y, x in $P_{[A]}$ are such that $\rho_{[\alpha]}(x) = \rho_{[\alpha]}(y)$, for all $\alpha \in \mathcal{A}$, where we write $\rho_{[\alpha]}$ for $\rho^{[A]}_{[A_\alpha]}$. Let x, respectively y in $\varinjlim_{\alpha \in A} P(a)$ be represented by $\{x_a; a \in A, x_a \in P(a)\}$ respectively $\{y_a; a \in A, y_a \in P(a)\}$. Since for all $\alpha \in \mathcal{A}$ we have $[A] \geq [A_\alpha]$, there are $a_\alpha \in A_\alpha, a_\alpha \leq a$ for a given $a \in A$. Fixing $a \in A$ and x_a, y_a in $P(a)$, $\rho^a_{a_\alpha}(x_a) = \rho^a_{a_\alpha}(y_a)$ for some $a_\alpha \in A_\alpha$ because we have that $\rho_{[\alpha]}(x) = \rho_{[\alpha]}(y)$. Now $\vee\{a_\alpha, \alpha \in \mathcal{A}\} \in \dot{\vee} A_\alpha$ and $A \sim \dot{\vee} A_\alpha$, so we may replace the a_α by $b_\alpha \in A_\alpha$ such that $b_\alpha \leq a_\alpha$ and there are a' and a'' in A such that $a' \leq \vee\{b_\alpha, \alpha \in \mathcal{A}\} \leq a''$. Then pick $a_1 \in A$ such that $a_1 \leq a$ and $a_1 \leq a''$ (A is directed). Again from $\dot{\vee} A_\alpha \sim A$ it follows that we may find $b'_\alpha, \alpha \in \mathcal{A}, b'_\alpha \in \mathcal{A}, b'_\alpha \in A_\alpha$ such that $b'_\alpha \leq \gamma_\alpha$ and $\vee\{b'_\alpha, \alpha \in \mathcal{A}\} \in \dot{\vee} A_\alpha$ is such that $\vee\{b'_\alpha, \alpha \in \mathcal{A}\} \leq a_1 \leq a$. Similarly we may find $b''_\alpha \in A_\alpha, b''_\alpha < b_\alpha$ for all $\alpha \in \mathcal{A}$, such that there exist a_2 and a'_2 in A such that $a'_2 \leq \vee b'_\alpha \leq a_2$, but $\vee b''_\alpha \leq \vee b'_\alpha \leq a_1 \leq a$. For convenience, rewrite b_α for b''_α and put $b = \vee\{b_\alpha, \alpha \in \mathcal{A}\}$. Since P is separated and $\rho^b_{b_\alpha}\rho^a_b(x_a) = \rho^b_{b_\alpha}\rho^a_b(y_a)$ for all $\alpha \in \mathcal{A}$, it follows that $\rho^a_b(x_a) = \rho^a_b(y_a)$. Since we may choose $a'_2 \in A$ such that $a'_2 \leq b$, we get $\rho^a_{a'_2}(x_a) = \rho^a_{a'_2}(y_a)$; thus the classes of x_a and y_a coincide in $\varinjlim_{a \in A} P(a)$, i.e. $x = y$ follows as desired.

ii. Start from a cover $[A] = [A_1] \vee \cdots \vee [A_n]$ in $C(\Lambda)$ and suppose there is given a set $x^\alpha \in P_{[A_\alpha]}, \alpha = 1, \ldots, n$, such that, writing $[C]$ for either $[A_\beta] \wedge [A_\alpha]$ or $[A_\alpha] \wedge [A_\beta]$ we have the gluing condition:

$$(*) \qquad \rho^{[A_\alpha]}_{[C]}(x^\alpha) = \rho^{[A_\beta]}_{[C]}(x^\beta), \quad \text{for } \alpha, \beta \in \{1, \ldots, n\}$$

Let x^α, respectively x^β, be represented by $x^\alpha_{a_\alpha}$, respectively $x^\beta_{a_\beta}$ for $a_\alpha \in A_\alpha$, respectively $a_\beta \in A_\beta$. Since $[A_\alpha] \wedge [A_\beta] = [A_\alpha \dot{\wedge} A_\beta]$, condition (*) yields:

$$(\bullet) \qquad \rho^{a_\alpha}_\gamma(x^\alpha_{a_\alpha}) = \rho^{a_\beta}_\gamma(x^\beta_{a_\beta}) \quad \text{for some } \gamma = a'_\alpha \wedge a'_\beta$$

with $a'_\alpha \in A_\alpha, a'_\beta \in A_\beta$. In a similar way:

$$(\bullet\bullet) \qquad \rho^{a'_\alpha}_{\gamma'}(x^\alpha_{a_\alpha}) = \rho^{a_\beta}_{\gamma'}(x^\beta_{a_\beta}) \quad \text{for some } \gamma' = a''_\beta \wedge a''_\alpha$$

with $a''_\alpha \in A_\alpha, a''_\beta \in A_\beta$. Take $b_\alpha \le a'_\alpha, a''_\alpha$ and $b_\beta \le a'_\beta, a''_\beta$, put $\gamma_{\alpha\beta} = b_\alpha \wedge b_\beta, \gamma_{\beta\alpha} = b_\beta \wedge b_\alpha$ in $A_\alpha \dot{\wedge} A_\beta$, respectively $A_\beta \dot{\wedge} A_\alpha$. Since $\gamma_{\alpha\beta} \le \gamma$ the equality ($\bullet$) also holds with γ replaced by $\gamma_{\alpha\beta}$ and ($\bullet\bullet$) holds with γ' replaced by $\gamma_{\beta\alpha}$. Since we considered a finite cover we may repeat the foregoing argument for all pairs A_α, A_β until we obtain a set $\{a_1, \ldots, a_n\}$ and $\gamma_{\alpha\beta} = a_\alpha \wedge a_\beta$ in $A_\alpha \dot{\wedge} A_\beta$, $\gamma_{\beta\alpha} = a_\beta \wedge a_\alpha$ in $A_\beta \dot{\wedge} A_\alpha$ such that the equalities ($\bullet$) and ($\bullet\bullet$) hold for all α and β in $\{1, \ldots, n\}$ with respect to $\gamma_{\alpha\beta}$, respectively $\gamma_{\beta\alpha}$. Since P is a sheaf on Λ it follows that there exists a $z \in P(\tau)$, $\tau = a_1 \vee \ldots \vee a_n$ such that $\rho^\tau_{a_\alpha}(z) = x^\alpha_{a_\alpha}$ for all $\alpha \in \{1, \ldots, n\}$.

Now $\tau \in A_1 \dot{\vee} \ldots \dot{\vee} A_n$ defines an element of the class $[A] = [A_1] \vee \ldots \vee [A_n]$ and z defines an element, say x, in $P_{[A]}$. Now both elements $\rho^{[A]}_{[A_\alpha]}(x)$ and x^α in $P_{[A_\alpha]}$ are represented by $\rho^\tau_{a_\alpha}(x_\tau)$, respectively $\rho^\tau_{a_\alpha}(z)$ for all $\alpha \in \mathcal{A} = \{1, \ldots, n\}$. Since the latter are the same and $P_{[\ \ ]}$ is separated in view of i, it follows that $x_\tau = z$ and x is the unique element of $P_{[A]}$ with the desired restrictions in each $P_{[A_\alpha]}$, $\alpha = 1, \ldots, n$. □

Observe that a separated presheaf P on Λ does not necessarily induce a separated presheaf on $SL(\Lambda)$ with respect to the operation $\dot{\wedge}$ in $\mathrm{id}_\wedge(\Lambda)$! We shall return to the sheaf theory in subsequent sections.

The notion of sheaf as we have used it so far presents a drawback; indeed if we start from a cover of $\lambda \in \Lambda$, say $\lambda = \lambda_1 \vee \ldots \vee \lambda_n$, and $\mu \le \lambda$ in Λ, then $\mu \ne (\mu \wedge \lambda_1) \vee \ldots \vee (\mu \wedge \lambda_n)$ in general. Even if Λ is a lattice but not distributive we meet this problem. Consequently, the gluing condition in the definition of a sheaf is too strong. Similar to the trick used in defining Grothendieck topologies we may restrict the covers used in the definition of a sheaf, and for example, only consider covers of λ induced by finite global covers in Λ. Let us write the more general definition in its functorial form; we do not demand that objects of $\underline{\mathcal{C}}$ are sets from now on, but we do assume that $\underline{\mathcal{C}}$ is a Grothendieck category. Consider a presheaf $P : \underline{\Lambda}^o \to \underline{\mathcal{C}}$. Obviously presheaves on Λ with values in $\underline{\mathcal{C}}$ together with presheaf morphisms make up a category $\mathcal{Q}(\Lambda, \underline{\mathcal{C}})$. For $\lambda \in \Lambda$ we let $\mathrm{Cov}_\Lambda(\lambda)$ consist of those covers for λ induced by a finite global cover of Λ. We make $\mathrm{Cov}_\Lambda(\lambda)$ into a category by defining $\mathcal{U} \to \mathcal{V}$ if $\mathcal{U} = (\lambda_i)_{i\in I}$, $\mathcal{V} = (\mu_j)_{j \in J}$ and there is a map $\varepsilon : I \to J$ such that $\lambda_i \le \mu_{\varepsilon(i)}$ for all $i \in I$.

Given P and $\lambda \in \Lambda$ we define a contravariant functor $[P, \lambda] : \mathrm{Cov}_\Lambda(\lambda) \to \underline{\mathcal{C}}$ as follows. First we have for $\mathcal{U} = (\lambda_i)_{i \in I}$ in $\mathrm{Cov}_\Lambda(\lambda)$ projection morphisms:

$$p_i : \prod_{i \in I} P(\lambda_i) \to P(\lambda_i),$$

then we have a morphism $j : P(\lambda) \to \prod_{i \in I} P(\lambda_i)$ such that $p_i j = \rho^\lambda_{\lambda_i}$. We also have morphisms:

$$p_l, q_l, p_r, q_r : \prod_{i \in P} P(\lambda_i) \rightrightarrows\!\!\rightrightarrows \prod_{(j,k) \in I \times I} P(\lambda_i \wedge \lambda_j)$$

where the (j,k)-component of p_l is $\rho^{\lambda_j}_{\lambda_j\wedge\lambda_k}$, of q_l is $\rho^{\lambda_k}_{\lambda_j\wedge\lambda_k}$, of p_r is $\rho^{\lambda_j}_{\lambda_k\wedge\lambda_j}$ and of q_r is $\rho^{\lambda_k}_{\lambda_k\wedge\lambda_j}$, corresponding to a diagram in $\underline{\mathcal{C}}$:

$$\begin{array}{ccc} P(\lambda_j) & \longrightarrow & P(\lambda_j \wedge \lambda_k) \\ & \times & \\ P(\lambda_k) & \longrightarrow & P(\lambda_k \wedge \lambda_j) \end{array}$$

Thus, in $\underline{\mathcal{C}}$, we obtain a diagram, for $(\lambda_i)_{i\in I}$ in $\mathrm{Cov}_\Lambda(\lambda)$:

$$(*) \qquad P(\lambda) \mapsto_j \prod_{i\in I} P(\lambda_i) \rightrightarrows \prod_{(j,k)\in I\times I} P(\lambda_j \wedge \lambda_k)$$

Definition 2.4
The presheaf P is a **sheaf** if and only if the diagrams (*) are equalizer diagrams. Now we put $[P,\lambda](\mathcal{U})$ equal to the kernel of

$$\prod_{i\in I} P(\lambda_i) \rightrightarrows \prod_{(j,k)\in I\times I} P(\lambda_j \wedge \lambda_k).$$

We can define $LP \in \mathcal{Q}(\Lambda, \underline{\mathcal{C}})$ by putting

$$LP : \Lambda^o \to \underline{\mathcal{C}}, \lambda \to \varinjlim_{\mathcal{U}\in \mathrm{Co}\Lambda \mathrm{J}_\Lambda(\lambda)} ([P,\lambda](\mathcal{U}))$$

Lemma 2.20
The notions of separated presheaf and sheaf are now defined with respect to the fixed type of covers. If P is a separated presheaf, then the canonical morphism $P \to LP$ in $\mathcal{Q}(\Lambda, \underline{\mathcal{C}})$ is a monomorphism and LP is a sheaf; hence the functor LL may be viewed as a sheafification functor.

Proof
Similar to the classical case. □

The foregoing lemma (in particular its proof) depends heavily on axiom A.10; therefore sheafification does not necessarily work well even for lattices. This can be seen in Section 2.7 for the lattice $L(H)$ of a Hilbert space H. However, it turns out that for sheaf theoretic applications the absence of A.10 may be compensated for by the existence of enough quantum points (cf. Section 2.3).

2.4.1 Project: Quantum Points and Sheaves

In the situation of Definition 2.3, suppose Λ has a weak quantum basis $\{[B_\alpha], \alpha \in \mathcal{A}\}$ and consider a separated presheaf on Λ, say P. Look at the separated presheaf $P_{[\Lambda]}$

on $C(\Lambda)$. We have $\vee\{[B_\alpha], \alpha \in \mathcal{A}\} = 1$ and for all α, β in $\mathcal{A}$, $[B_\alpha] \cap [B_\beta] = 0$; up to self-intersections the $[B_\alpha]$ allows us to reconstruct $[\lambda] \in C(\Lambda)$ with $\lambda \in \Lambda$. Starting from the product $\coprod_{\alpha \in \mathcal{A}} P_{[B_\alpha]}$, of all stalks at the $[B_\alpha]$, imitate the construction of the étale space of a separated presheaf $P_{[\ \ ]}$ in two ways; first by putting restrictions on P such that $P_{[B_\alpha]} = P_{[B_\alpha]^n]}$, and then more generally by replacing the product above by $\coprod_{\alpha \in \mathcal{A}} P_{[B_\alpha]^{n_\alpha}}$ where $[B_\alpha]^{n_\alpha}$ is the idempotent dominated by $[B_\alpha]$. In case radical elements exist, it is quite harmless to assume that the stalk representing a radical element is the global $P_{[0]}$. Study the (partial) sheafification results that follow. Connect this to Projects 2.3.1.1 and 2.3.1.2 and ii. Observe that for the example $L(H)$, introduced in Section 2.7, this leads to the classical sheafification on the Stone space (see also later for generalizations of the Stone space). Observe that $x, y \in P_{[B_\alpha]}$ such that $x \neq y$ but $\rho^{\alpha}_{\alpha^m}(x) = \rho^{\alpha}_{\alpha^m}(y)$ in the $P_{[B_\alpha]^m}$ may exist. So the étale space in the second approach will be a quotient of $\coprod_{\alpha \in \mathcal{A}} P_{[B_\alpha]}$ identifying elements that are restricted to the same element after n self-intersections for some $n \in \mathbb{N}$.

2.5 Noncommutative Grothendieck Topologies

In the definition of noncommutative topology we have included some conditions about covers, such as A.10. When trying to fit the noncommutative topologies into the framework of Grothendieck topologies this fact will pay off. First recall that a Grothedieck topology is a category $\underline{\mathcal{C}}$ such that for every object x of $\underline{\mathcal{C}}$ a set $\text{cov}(x)$, consisting of subsets of morphisms with target x, is given such that the following conditions are satisfied:

G.1. $\{x \to x\} \in \text{cov}(x)$.

G.2. If $\{x_i \to x, i \in \mathcal{J}\} \in \text{cov}(x)$ and $\{x_{ij} \to x_i, j \in \mathcal{J}'\} \in \text{cov}(x_i)$ for all $i \in \mathcal{I}$, then $\{x_{ij} \to x, i \in \mathcal{J}, j \in \mathcal{J}'\} \in \text{cov}(x)$, where $x_{ij} \to x$ is obtained from the $x_{ij} \to x_j \to x$.

G.3. If $\{x_i \to x, i \in \mathcal{J}\} \in \text{cov}(x)$ and $x' \to x$ in $\mathcal{C}$, then there exists a pull-back $x' \times_x x_i$ in $\underline{\mathcal{C}}$ such that $\{x' \times_x x_i \to x', i \in \mathcal{J}\} \in \text{cov}(x')$, which is called the **fibre product over** x.

Now look at a noncommutative topology, Λ say. For the objects of $\underline{\mathcal{C}}$ there is little choice but to take $\lambda \in \Lambda$. Since we have to be able to induce covers, we cannot just let any relation $\lambda \leq \mu$ be a morhphism $\lambda \to \mu$ in $\underline{\mathcal{C}}$. A first idea could be to allow only focused relations $\lambda \leq \mu$ (as defined after Section 2.2.1.); however, this leads to problems such as if $\lambda_1 \vee \ldots \vee \lambda_n$ is a global cover and $x' \leq x$ is focused then we do get a diagram:

$$\begin{array}{ccc} x \wedge \lambda_i & \longrightarrow & x \\ \uparrow & & \uparrow \\ x' \wedge \lambda_i & \longrightarrow & x' \end{array}$$

in the category $\underline{C}$ but $x' \wedge \lambda' \leq x \wedge \lambda_i$ is not necessarily focused; that is, it may happen that $(x' \wedge \lambda_i) \wedge (x \wedge \lambda_i) \neq x' \wedge \lambda_i$. We may solve the problems by looking at generic relations. A relation $\lambda \leq \mu$ in Λ is said to be **generic** if it is a consequence of the axioms of a noncommutative topology; that is $a \wedge b \leq a$ is a generic relation. When a and b are idempotent in λ, any relation that is just given as $a \leq b$ is not viewed as generic; however, if $b = a \vee c$, then $a \leq b$ is viewed as generic.

Now if $1 = \lambda_1 \vee \ldots \vee \lambda_n$ is a global cover and $x' \leq x$ is generic, then $x' \wedge \lambda_i \leq x \wedge \lambda_i$ is generic too, so if we define $\underline{C}^g$ to be the category with objects $\lambda \in \Lambda$ and generic relations for the morphisms (with $x = x$ representing 1_x), then the diagram above is a diagram of morphisms in $\underline{C}^g$. A new problem appears, for example, for $x' \leq x$ and $1 = \lambda_1 \vee \ldots \vee \lambda_n$ as above; for given morphisms $t \to x \wedge \lambda_i, t \to x'$ we do not find a generic relation $t \leq x' \wedge \lambda_i$. But from the philosophy of patterns we may learn that it is not natural to ask for $t \to x' \wedge \lambda_i$; we should ask for some morphism between elements having a similar pattern. This is at the basis of the following definition.

Definition 2.5
A category $\underline{C}$ with $\mathrm{cov}(x)$ defined for every object x of $\underline{C}$ as before is said to be a **noncommutative Grothendieck topology** if the following conditions hold:

G.1. and **G.2.** as before, and the new condition:

G.3. **nc** for given $x' \to x$ and $\{x_i \to x, i \in \mathcal{J}\} \in \mathrm{cov}(x)$ there is a cover $\{x' \times_x x \to x', i \in \mathcal{J}\} \in \mathrm{cov}(x')$ satisfying the following pull-back property: for $s \to x_i, s \to x'$ and $t \to x_i, t \to x'$ there exist $s \wedge t \to x' \times_x x_i$ and $t \wedge s \to x' \times_x x_i$ fitting in a commutative diagram:

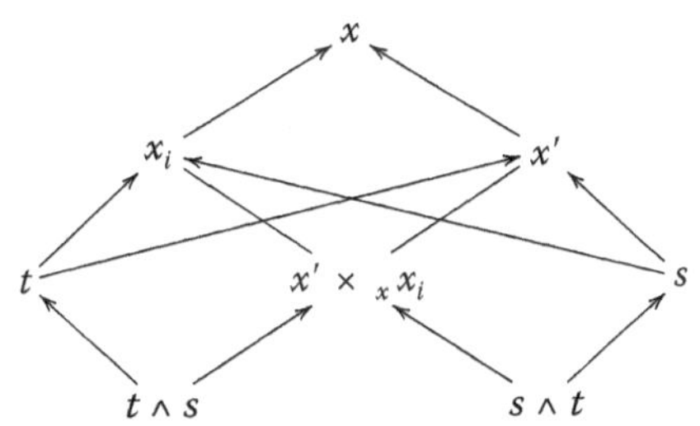

The particular case when $s = t$ yields:

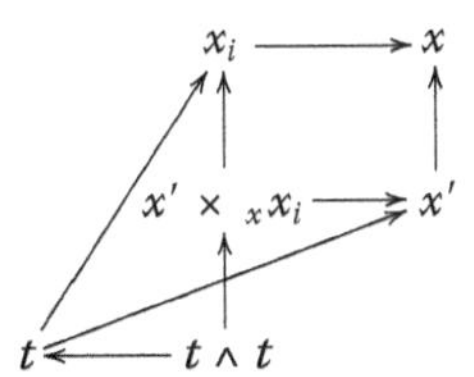

Clearly when t is idempotent, then the diagram reduces to the usual pull-back diagram in **G.3.** Observe that the noncommutative version of a Grothendieck topology is obtained by a complete symmetrization of the classical definition.

Theorem 2.2
Let X be either $\Lambda, C(\Lambda), \tau$, so certainly a noncommutative topology and let $\underline{\mathcal{C}}^g$ be constructed on X with respect to generic relations as before Definition 2.5. For the covers of x in X we take $\text{cov}_l(x) = \{x \wedge \lambda_i \to x, \{\lambda_1, \ldots, \lambda_n\}$ a global cover$\}$. Then X becomes a noncommutative Grothendieck topology.

Proof
Look at a cover $\{x_i = x \wedge \lambda_i \to x, i = 1, \ldots, n\} \in \text{cov}_l(x)$ and $x' \to x$. Put $x' \times_x x_i = x' \wedge \lambda_i$, for $i = 1, \ldots, n$. Since $x' \to x$, $x' \leq x$ is generic; thus $x' \wedge \lambda_i \leq x \wedge \lambda_i$ is generic too. We obtain the following diagrams in $\underline{\mathcal{C}}^g$:

$$\begin{array}{ccc} x \wedge \lambda_i & \longrightarrow & x \\ \uparrow & & \uparrow \\ x' \times_x x_i = x' \wedge \lambda_i & \longrightarrow & x' \end{array}$$

Suppose $s \to x \wedge \lambda_i$, $s \to x'$ and $t \to x \wedge \lambda_i$, $t \to x'$ are given. Since $x \wedge \lambda_i \leq \lambda_i$ is generic by definition and $s \to x \wedge \lambda_i, t \to x \wedge \lambda_i$ are morphisms in $\underline{\mathcal{C}}^g$, it follows that $s \leq x \wedge \lambda_i \leq \lambda_i$ and $t \leq x \wedge \lambda_i \leq \lambda_i$ yield generic relations $s \leq \lambda_i, t \leq \lambda_i$. Since $s \leq \lambda_i$ and $s \leq x'$ are generic and similarly $t \leq \lambda_i, t \leq x'$ are generic, it follows that $s \wedge t \leq x' \wedge \lambda_i$ and $t \wedge s \leq x' \wedge \lambda_i$ are generic. Hence we arrive at the morphism in $\underline{\mathcal{C}}^g$, $s \wedge t \to x' \wedge \lambda_i = x' \times_x x_i$ and $t \wedge s \underset{\text{nc}}{\longrightarrow} x' \wedge \lambda_i = x' \times_x x_i$, fitting nicely in the diagram defining the condition **G.3** (**G.1** and **G.2** are obvious), so X (as $\underline{\mathcal{C}}^g$) is a noncommutative Grothendieck topology. □

Remark 2.1

1. Using cov_l we refer to X with this structure as the **left topology** of Λ (or T, or $C(\Lambda)$, τ); in a similar way the **right topology** can be defined.
2. The advantage of using generic relations is obvious: the generic relations are recognizable on sight; they transfer well from Λ to $C(\Lambda)$ and τ. The fact that not every generic relation needs to be focused does not interfere with that.
3. When viewing $\Lambda, C(\Lambda), \tau$ with the noncommutative Grothendieck topology structure, we shall write $\underline{\Lambda}^g$, $\underline{C(\Lambda)}^g$, $\underline{\tau}^g$, respectively.

Observation 2.3
A cover $\lambda = \lambda_1 \vee \ldots \vee \lambda_n$ in X (as before) is automatically a **generic cover** because $\lambda_i \leq \lambda$ is generic for $i = 1, \ldots, n$.

Now it is clear how **presheaves, separated presheaves, or sheaves are defined over** X with respect to **generic covers**. Indeed the operations appearing in the separateness and the sheaf property are compatible with the restriction to generic relations and generic covers. When passing from Λ to $C(\Lambda)$ or τ we may look at a stronger restriction on the covers (much like the strong idempotents replacing the idempotents). For [A], [B] in $C(\Lambda)$ the relation $[A] \leq [B]$ is **generically defined** if for all $b \in B$ there exists an $a \in A$ such that $a \leq b$ is generic. A cover $[A] = \cup_\alpha [A_\alpha]$ is generically defined if $[A_\alpha] \leq [A]$ is generically defined for all α.

Lemma 2.21
For $C(\Lambda)$ consider the category $C(\Lambda)^{gd}$ consisting of the objects $[A] \in C(\Lambda)$ with generically defined relations for the morphisms; then $\underline{C(\Lambda)}^{gd}$ is a noncommutative Grothendieck topology.

Proof
The covers used are generically defined covers; **G.1** and **G.2** are obvious. For G_3^{nc} the proof in Theorem 2.2 remains valid if we verify that a generically defined relation $a \leq b$ yields a generically defined $x \wedge a \leq x \wedge b$ for any x. Put $a = [A], b = [B]$ and $x = [C]$ in $C(\Lambda)$. Since $[A] \leq [B]$ is generically defined, we find for every $b \in B$ some $a \in A$ with $a \leq b$ generic. For any $x \in C$ we then have a generic relation $a \wedge c \leq b \wedge c$; hence for every $b \wedge c \in B \dot{\wedge} C$ there is an $a \wedge c \in A \dot{\wedge} C$ such that $a \wedge c \leq b \wedge c$ is generic. Consequently, $[a] \wedge [C] \leq [B] \wedge [C]$ is generically defined. □

Proposition 2.9
If P is a presheaf on $\underline{\Lambda}^g$, then $P_{[\,]}$ is a presheaf on $\underline{C(\Lambda)}^{gd}$. Moreover:

i. *If P is separated, then $P_{[\,]}$ is separated.*
ii. *If P is a sheaf, then $P_{[\,]}$ is a sheaf on $\underline{C(\Lambda)}^{gd}$*

Proof
That $P_{[\,]}$ is now a presheaf on $\underline{C(\Lambda)}^{gd}$ follows from the (proof of) Lemma 2.19.

i. The same proof as in Theorem 2.1 i holds still if one verifies at each step that the relations may be chosen to be generic; then the only operations occuring are $\wedge$ or $\vee$ and these do not change the generic character.
ii. Same as in i, checking along the lines of Theorem 2.1 ii. □

Corollary 2.7
The canonical inclusion $\Lambda \hookrightarrow C(\Lambda), \lambda \mapsto [\lambda]$ defines a faithful $\underline{\Lambda}^g \to \underline{C(\Lambda)}^{gd}$. Indeed a generic relation $\lambda \leq \mu$ does define a generically defined $[\lambda] \leq [\mu]$. The restriction of $P_{[\,]}$ to $\underline{\Lambda}^g$ is P.

Corollary 2.8
The pattern topology $\tau \subset C(\Lambda)$ is also a noncommutative Grothendieck topology, written $\underline{\tau}^g$. Defining $\underline{\tau}^{gd}$ in the same way as $\underline{C(\Lambda)}^{gd}$, this also defines another noncommutative Grothendieck topology and a faithful functor $\underline{\tau}^{gd} \to \underline{C(\Lambda)}^{gd}$. Presheaves on $\underline{C(\Lambda)}^{gd}$ restrict to $\underline{\tau}^{gd}$ in such a way that separateness and the sheaf property are respected.

2.5.1 Warning

If a and b are idempotents in a topology of virtual opens Λ, then the fact that $a \leq b$ is generic in Λ need not imply that $a \leq b$ is generic in $SL(\Lambda)$! For example if

$a = (b \vee c) \wedge (d \vee c)$, then $c \wedge c \leq a$ is generic in Λ but c and d are not even idempotent, so we must relate a and c via something like $(b \vee c) \wedge (d \vee c)$. This **bad behavior of *genericity* with respect to the commutative shadow** prompts us to work with sheaves on the lattice-type noncommutative topology, avoiding the Grothendieck topologies here for the moment, at least when one aims to relate Λ and $SL(\Lambda)$. A good theory of sheaves over noncommutative Grothendieck topologies probably has to be developed in connection with noncommutative topos theory. If one studies noncommutative Grothendieck topologies without reference to another noncommutative topology from which it stems, the sheaf theory is of independent interest.

2.5.2 Projects

2.5.2.1 A Noncommutative Topos Theory

What structure fits the philosophical equation:

$$\frac{\text{locales}}{\text{quantales}} = \frac{\text{topoi}}{?}$$

The answer should lead to some version of noncommutative topos theory; however, let me point out that it is not clear to this author whether the above question is the correct one to put forward; quantales are too tightly related to C^*-algebras to obtain the right level of generality perhaps; nevertheless, the search for a noncommutative topos is worthwhile in its own right. One easily finds that the first main problem is to circumvent the notion of subobject classifier. A first generalized theory may be constructed by maintaining the "up to self-intersections" philosophy; a second approach may be to allow a **family** of "subobject classifiers" defined in a suitable way. It is clear that noncommutative topology reflects a kind of ordered logic; the ordering reflects the fact that "$x \in A$" and "$x \in B$" cannot be realized at the same time. We have not yet tried to write down a foundation for such ordered logic, but certainly a notion of noncommutative topos would fit perfectly in this. We leave the development of noncommutative topos theory as a project here; nevertheless we do know the main examples to be covered by such a theory: sheaves over a noncommutative Grothendieck topology, in particular those constructed on a generalized Stone space or even more specifically on a noncommutative Grothendieck topology constructed from a quantum topology (because of the presence of suitable sheafification techniques). Detail on topos theory may be found in R. Goldblatt's [15].

2.5.2.2 Noncommutative Probability (and Measure) Theory

The mathematician may be completely satisfied with the foundation of probability theory knowing that it means exactly what it means and no more! There is more than one question to be raised concerning certain applications in the real world or what passes for that frequently. You can only throw the dice until they crack! The certainty that some event in a given selection must happen (and to associate to this a number related to a total number of possible events in the selection) is rather ill founded. The notion of the "time necessary for some events to actually happen" is neglected, and at the level of theoretical foundation probability is based on the membership relations of set theory to be time or ordering independent. The idea of a noncommutative space is not consistent with such a probability theory, so it must be constructed ab initio in a noncommutative way too. An intersection $A \wedge B$ in a noncommutative space should be related to a conditional type of probability in the sense that the probability for $\lambda \leq A \wedge B$ is expressed by $p(\lambda, A)p(A, B)p(\lambda, B)$ where $p(A, B)$ expresses the probability correction for A before B. Since a σ-algebra for some noncommutative topology may be defined in a straightforward way, noncommutative versions of Borel-stems, etc. are not hard to develop. Probabilities could be taken in the free semigroup (generalizing $\mathbb{N}$) over the set Λ, making multiplication of probabilities formal and noncommutative but perhaps also unnatural. The approach suggested above, that is using the conditional approach retains a classical flavor, again the self-intersection introduces new phenomena, for example, $p(A, A)$ may be nontrivial, that need to be fully integrated in the theory.

2.5.2.3 Covers and Cohomology Theories

Traditionally the idea of Grothendieck topology and the extra abstraction in the notion of cover allows us to introduce new interesting (co)homology theories; recall the use of étale covers and étale cohomology. In [46] and [50] we did a similar experiment with respect to Čech cohomology, which led to some interesting results in the algebraic theory of noncommutative geometry. For example, a result of L. Le Bruyn concerning the moduli space of left ideals in Weyl algebras has been reduced to a fairly straightforward calculation of Čech cohomology on the noncommutative site created from the noncommutative topology phrased in terms of Ore sets in the algebra considered. Similarly, V. Ginzburg and A. Berest have used the same technique in another situation.

Of course one should be tempted to develop a noncommutative étale cohomology or more cohomology theories with respect to other types of covers. Since examples of noncommutative topologies may be constructed in a completely functorial way (see also Chapter 4) one may start a theory from the consideration of covers by separable functors in the sense of Section 1.3.

2.5.2.4 The Derived Imperative

For compactness sake derived categories and derived functors have not been introduced in these notes; consequently, our sheaves are not perverse. Clearly, the latter are popular topics nowadays and they also provide strong methods of analysis, for example, in connection with rings of differential operators, Riemann-Hilbert correspondence, and so forth. It is a promising idea to combine the techniques of derived categories, perverse sheaves, and so forth with the noncommutative topology point of view. We say no more about this here.

2.6 The Fundamental Examples I: Torsion Theories

We need to recall the basic facts about torsion theory. What we say will be valid for an abelian category that is assumed to be complete, co-complete, and locally small, but we shall restrict attention to Grothendieck categories for convenience.

Let $\underline{\mathcal{C}}$ be a Grothendieck category. A **preradical** ρ of $\underline{\mathcal{C}}$ is just a subfunctor of the identity functor. The class of preradicals of $\mathcal{C}$, $\mathcal{Q}$ say, is partially ordered by $\rho_1 \leq \rho_2$ if and only if $\rho_1(C) \subset \rho_2(C)$ for all objects C of $\underline{\mathcal{C}}$. Any family of preradicals $\{\rho_\alpha, \alpha \in \mathcal{A}\}$ has at least an upperbound $\vee\rho_\alpha$ and a greatest lower bound $\wedge\rho_\alpha$ defined in the obvious way. Consequently, $\mathcal{Q}(\underline{\mathcal{C}})$ is a complete lattice with respect to $\wedge$ and $\vee$. For preradicals ρ_1 and ρ_2 on $\underline{\mathcal{C}}$ we may also define $\rho_1\rho_2$ by putting $\rho_1\rho_2(C) = \rho_1(\rho_2(C))$ for all $c \in \underline{\mathcal{C}}$; we define $\rho_1 : \rho_2$ by taking $(\rho_1 : \rho_2)(C)$ for $C \in \underline{\mathcal{C}}$, to be the subobject of C for which we have $(\rho_1 : \rho_2)(C)/\rho_1(C) = \rho_2(C/\rho_1(C))$.

Definition 2.6
A preradical ρ such that $\rho\rho = \rho$ is said to be **idempotent**. We say that a preradical is a **radical** if $(\rho : \rho) = \rho$; in other words ρ is radical if $\rho((C)/\rho(C)) = 0$ for $C \in \underline{\mathcal{C}}$. To a preradical ρ of $\underline{\mathcal{C}}$ we may associate a preradical ρ^{-1} of $\underline{\mathcal{C}}^o$ by defining $\rho^{-1}(X) = X/\rho(X)$; we call ρ^{-1} the **dual preradical** of ρ.

It is easy to establish that if ρ is idempotent, respectively radical, then ρ^{-1} is respectively radical, idempotent. A preradical ρ of $\underline{\mathcal{C}}$ gives rise to two classes of objects in $\underline{\mathcal{C}}$:

$$\mathcal{F}_\rho = \{C \in \underline{\mathcal{C}}, \rho(C) = 0\} \text{ (pretorsion free class)}$$

$$\mathcal{T}_\rho = \{C \in \underline{\mathcal{C}}, \rho(C) = C\} \text{ (pretorsion class)}$$

Observe that $\mathcal{F}_\rho = \mathcal{T}_{\rho^{-1}}$ objectwise.

The relation between ρ and these classes is well summarized in the following.

Theorem 2.3
With notation and conventions as above:

i. *$\mathcal{T}_\rho$ is closed under quotient objects and coproducts; $\mathcal{F}_\rho$ is closed under subobjects and products.*
ii. *If $T \in \mathcal{T}_\rho$ and $F \in \mathcal{F}_\rho$, then* $\mathrm{Hom}_{\underline{\mathcal{C}}}(\mathrm{T}, \mathrm{F}) = 0$.
iii. *Idempotent preradicals of $\underline{\mathcal{C}}$ correspond bijectively to pretorsion classes of objects of $\underline{\mathcal{C}}$, that is, classes that are closed under quotient and coproducts.*
iv. *Radicals of $\underline{\mathcal{C}}$ correspond bijectively to pretorsion-free classes of objects of $\underline{\mathcal{C}}$, that is, classes that are closed under subobjects and products.*
v. *For every $\rho \in \mathcal{Q}(\underline{\mathcal{C}})$ there exists a largest idempotent preradical $\rho^o \leq \rho$ and a smallest radical $\rho^c \geq \rho$.*

Proof

i. Let us establish the first claim; the second follows by duality. That $\mathcal{T}_\rho$ is closed under quotient objects is easily seen. Look at a family $\{C_\alpha, \alpha \in \mathcal{A}\}$ of objects in $\underline{\mathcal{C}}$ and in $\mathcal{T}_\rho$. Because $\rho(C_\alpha) = C_\alpha$ for all $\alpha \in \mathcal{A}$, the C_α map under the canonical monomorphism $C_\alpha \to \oplus_\alpha C_\alpha$ into $\rho(\oplus_\alpha C_\alpha)$. The universal property of $\oplus$ then leads to the conclusion that $\rho(\oplus C_2) = \oplus C_\alpha$; that is, $\oplus C_\alpha$ is in $\mathcal{T}_\rho$ too.

ii. If $f : T \to F$ is a nonzero morphism, then Imf is in $\mathcal{T}_\rho$ because of i, but since it is also a subobject of F, Imf $= 0$ follows, so no nonzero f can exist.

iii. Consider a pretorsion class $\mathcal{T}$. To an arbitrary object C of $\underline{\mathcal{C}}$ we associate $t(C) \in \underline{\mathcal{C}}$ by considering $t(C)$ to be the sum of all subobjects of C that are objects of $\mathcal{T}$; then $t(C) \in \mathcal{T}$ because $\mathcal{T}$ is closed under coproducts and quotient objects. It is easy to see that t is an idempotent preradical of $\underline{\mathcal{C}}$. It is also clear that $\mathcal{T}_t = \mathcal{T}$. Now if we start with an idempotent preradical ρ, $\mathcal{T}_\rho$ is a pretorsion class in $\underline{\mathcal{C}}$ because of i; the idempotent preradical t_ρ associates to C in $\underline{\mathcal{C}}$ the largest subobject C' of C such that $\rho(C') = C'$, but that is exactly $\rho(C)$; hence $\rho = t_\rho$.

iv. Dual to iii.

v. Starting from a preradical ρ of $\underline{\mathcal{C}}$ we define a pretorsion class $\mathcal{T}_\rho$ and an idempotent preradical t_ρ defined (see proof of iii.) by taking $t_\rho(C)$ to be the largest subobject of C, say C', such that $\rho(C') = C'$. Therefore $t_\rho(C) \subset \rho(C)$ for all $C \in \underline{\mathcal{C}}$, that is, $t_\rho \le \rho$ in $\mathcal{Q}(\underline{\mathcal{C}})$, and t_ρ is clearly the largest idempotent preradical of $\underline{\mathcal{C}}$ with this property, so we may put $\rho^o = t_\rho$. The second claim, concerning ρ^o, follows by duality. □

A **closure operator** on a complete lattice L is a map $(-)^c : L \to L, \lambda \mapsto \lambda^c$, satisfying the following:

c.1 If $\lambda \le \mu$ in L then $\lambda^c \le \mu^c$

c.2 For $\lambda \in L$ we have that $\lambda \le \lambda^c$

c.3 For $\lambda \in L$ we have that $(\lambda^c)^c = \lambda^c$.

The set of closed elements of L, that is, those λ for which we have $\lambda = \lambda^c$, forms a complete lattice L^c with respect to $\le$ and $\wedge$ as in L but with $\overline{\vee}$ defined by $\overline{\vee}\lambda_\alpha = (\vee\lambda_\alpha)^c$.

Observation 2.4
If L is a complete modular lattice with closure operator $(-)^c$ satisfying $(\lambda \wedge \mu)^c = \lambda^c \wedge \mu^c$, then $L^c, \le, \wedge, \overline{\vee}$ is a complete modular lattice too.

Proof
If λ, μ, γ are closed elements with $\lambda \le \mu$ then:

$$\begin{aligned} \mu \wedge (\gamma \overline{\vee} \lambda) = \mu^c \wedge (\gamma \vee \lambda)^c &= (\mu \wedge (\gamma \vee \lambda))^c \text{ (by the assumption)} \\ &= (\lambda \vee (\gamma \wedge \mu))^c \text{ (L is modular)} \\ &= \lambda \overline{\vee} (\gamma \wedge \mu) \end{aligned}$$

□

Now considering $\mathcal{Q}(C)$, it is clear that $(-)^c : \mathcal{Q}(\underline{\mathcal{C}}) \to \mathcal{Q}(\underline{\mathcal{C}})$ is a closure operator. Therefore the foregoing observation implies that the idempotent radicals, respectively the radicals, form a complete lattice. Indeed $\mathcal{Q}(R)$ satisfies the condition $(\lambda \wedge \mu)^c = \lambda^c \wedge \mu^c$ as is easily verified and the second statement follows by duality.

Proposition 2.10

i. If ρ is idempotent, then so is ρ^c.

ii. If ρ is a radical, then so is ρ^c.

Proof
It suffices to establish i; then ii follows by duality. From $\rho\rho = \rho$ we have to establish that $\rho^c\rho^c = \rho^c$. Now for $C \in \underline{\mathcal{C}}$ $\rho^c(C)$ is the smallest subobject of C such that $\rho(C/\rho^c(C)) = 0$, and $\rho^c\rho^c(C)$ is the smallest subobject of $\rho^c(C)$ such that $\rho(\rho^c(C)/\rho^c\rho^c(C)) = 0$. Look at $C \supset \rho^c(C) \supset \rho^c\rho^c(C)$. If $T \subset C$ is such that $T/\rho^c\rho^c(C)$ is in $\mathcal{T}_\rho$, then also $T + \rho^c(C)/\rho^c(C) \in \mathcal{T}_\rho$ and therefore $T \subset \rho^c(C)$. The latter yields that $T/\rho^c\rho^c(C)$ is in $\mathcal{T}_\rho$ and thus $T = \rho^c\rho^c(C)$. Therefore we arrive at $C/\rho^c\rho^c(C) \in \mathcal{F}_\rho$. On the other hand, if $D \subset \rho^c\rho^c(C)$ is such that C/D is in $\mathcal{F}_{\rho\rho} = \mathcal{F}$, then $(\rho\rho)^c(C) \subset D$ and then $\rho^c(C) \subset D$, but this entails that $\rho^c(C) = \rho^c\rho^c(C)$. □

Proposition 2.11
For $\rho \in \mathcal{Q}(\underline{\mathcal{C}})$, the following are equivalent:

i. ρ is left exact.

ii. For every subset D of C in $\underline{\mathcal{C}}$, $\rho(D) = \rho(C) \cap D$.

iii. ρ is idempotent and $\mathcal{T}_\rho$ is closed under subobjects.

Proof
Easy enough to be left as an exercise. □

A pretorsion class closed under subobjects is said to be **hereditary**; hence, by the proposition, hereditary pretorsion classes correspond bijectively to left exact preradicals.

Note that the operation $\rho_1\rho_2$ in $\mathcal{Q}(\underline{\mathcal{C}})$ is noncommutative. Note also that $(\rho\tau)^{-1}$ and $\tau^{-1}\rho^{-1}$ (in $\underline{\mathcal{C}}^o$) are different preradicals. Whereas duality works perfectly when $\vee$ and $\wedge$ are being considered, it breaks down for the noncommutative operations. One may interpret this as if $\rho\tau$ tries to be a noncommutative intersection while $\tau^{-1}\rho^{-1}$ tries to be a noncommutative union. Phrasing this in $\Lambda(\underline{\mathcal{C}}) = \mathcal{Q}(\mathcal{C})^o$ we reobtain the possibility of using a commutative union stemming from $\wedge$ in $\mathcal{Q}(\underline{\mathcal{C}})$ and a noncommutative "intersection" (when viewed in $\mathcal{Q}(\underline{\mathcal{C}})^o$) stemming from the preradical product $\tau^{-1}\rho^{-1}$ for $\tau, \rho \in \mathcal{Q}(\underline{\mathcal{C}})$. We shall make this more precise in Section 3.1.

First we look now at those preradicals that are both idempotent and radical.

Definition 2.7
A torsion theory for $\underline{\mathcal{C}}$ is a pair $(\mathcal{T}, \mathcal{F})$ of classes of objects from $\underline{\mathcal{C}}$ such that:

tt1 For $T \in \mathcal{T}$, $F \in \mathcal{F}$, $\mathrm{Hom}_{\underline{\mathcal{C}}}(\mathrm{T}, \mathrm{F}) = 0$.

tt2 If $\mathrm{Hom}_{\underline{\mathcal{C}}}(\mathrm{C}, \mathrm{F}) = 0$ for all $F \in \mathcal{F}$ then $C \in \mathcal{T}$.

tt3 If $\mathrm{Hom}_{\mathcal{C}}(\mathrm{T}, \mathrm{C}) = 0$ for $T \in \mathcal{T}$ then $C \in \mathcal{F}$.

We say that $\mathcal{T}$ is a **torsion-class** of $\underline{\mathcal{C}}$ and its objects are $(\mathcal{T}, \mathcal{F})$-torsion objects of $\underline{\mathcal{C}}$, while $\mathcal{F}$ is the $(\mathcal{T}, \mathcal{F})$ **torsion-free class**.

A given class $\mathcal{M}$ in $\underline{\mathcal{C}}$ cogenerates a torsion theory $(\mathcal{T}, \mathcal{F})$, which is the smallest torsion-free class containing $\mathcal{M}$,

$$\mathcal{F}_{\mathcal{M}} = \{F \text{ in } \underline{\mathcal{C}}, \mathrm{Hom}_{\underline{\mathcal{C}}}(\mathrm{C}, \mathrm{F}) = 0 \text{ for all } \mathrm{C} \in \mathcal{M}\}$$
$$\mathcal{T}_{\mathcal{M}} = \{T \text{ in } \underline{\mathcal{C}}, \mathrm{Hom}_{\underline{\mathcal{C}}}(\mathrm{T}, \mathrm{F}) = 0 \text{ for all } \mathrm{F} \in \mathcal{F}_{\mathcal{M}}\}$$

Proposition 2.12: Characterization of Torsion Classes

a. For a class $\mathcal{T}$ of objects of $\underline{\mathcal{C}}$ the following are equivalent:

   i. $\mathcal{T}$ is the torsion class of some torsion theory.

   ii. $\mathcal{T}$ is closed under quotient objects, coproducts, and extensions; that is, for every exact sequence $0 \to C' \to C \to C'' \to 0$ in $\mathcal{C}$ with C' and C'' in $\mathcal{T}$, then $C \in \mathcal{T}$.

b. For a class $\mathcal{F}$ of objects of $\overline{\mathcal{C}}$ the following are equivalent:

   i. $\mathcal{F}$ is a torsion-free class for some torsion theory of $\underline{\mathcal{C}}$.

   ii. $\mathcal{F}$ is closed under subobjects, products, and extensions.

Proof

Old hat, see among others B. Stenström, *Rings of Quotients*, Springer Verlag, Heidelberg, 1975 [42].

A torsion theory $(\mathcal{T}, \mathcal{F})$ is in particular pretorsion, so it defines an idempotent preradical τ, which in view of the fact that $\mathcal{T}$ is closed under extensions, is easily seen to be a radical. Conversely, given an idempotent radical τ in $\mathcal{Q}(\underline{\mathcal{C}})$, it determines a torsion theory of $\underline{\mathcal{C}}$ by $\mathcal{F}_\tau = \{C \text{ in } \underline{\mathcal{C}}, \tau(C) = 0\}$, $\mathcal{T}_\tau = \{C \text{ in } \mathcal{C}, \tau(C) = C\}$. □

Proposition 2.13

Torsion theories correspond bijectively to idempotent radicals; if ρ is an idempotent preradical, then ρ^c is the idempotent radical corresponding to the torsion theory generated by $\mathcal{T}_\rho$.

*A torsion theory $(\mathcal{T}, \mathcal{F})$ is said to be **hereditary** if $\mathcal{T}$ is closed under submodules. In view of Proposition 2.11 and the foregoing proposition, it follows that there is a bijective correspondence between hereditary torsion theories and left exact radicals. If $\underline{\mathcal{C}}$ is a Grothendieck category with enough injectives, then we can characterize hereditary torsion theories by the fact that $(\mathcal{T}, \mathcal{F})$ is hereditary if and only if $\mathcal{F}$ is closed under injective envelopes.*

Proposition 2.14

If $\mathcal{M}$ is a class closed under subobjects and quotient objects (and $\underline{\mathcal{C}}$ is as mentioned above), then the torsion theory generated by $\mathcal{M}$ is hereditary.

Proof
Suppose F is torsion free and assume there is a nonzero $f : C \to E(F)$ for some $C \in \mathcal{M}$ where $E(F)$ is the injective envelope of F. Then Imf $\in \mathcal{M}$ and $F \cap$ Imf is a nonzero subobject of F belonging to $\mathcal{M}$ as the latter is closed under subobjects—a contradiction. □

Corollary 2.9

i. *If ρ is a left exact preradical, then ρ^c is also left exact.*

ii. *If ρ is a left exact preradical, then $\rho(C)$ is an essential subobject of $\rho^c(C)$; that is, for every subobject D, nonzero, in $\rho^c(C)$ we have that $D \cap \rho(C)$ is nonzero.*

Proof

i. By assumption $\mathcal{T}_\rho$ is a hereditary pretorsion class. The foregoing proposition (and the proof of iii in Theorem 2.3) yields that ρ^c is left exact.

ii. Suppose $D \cap \rho(C) = 0$. Then $\rho(D) = 0$; hence $\rho^c(D) = 0$, but $\rho^c(D) = \rho^c(C) \cap D$, and the latter is just D, thus $D = \rho(D) = 0$ follows. □

Note that a Grothendieck category with a generator has enough injectives.

Definition 2.8
A left exact idempotent radical is called a **kernel functor**. It is clear from the foregoing that kernel functors correspond bijectively to the hereditary torsion theories.

If κ denotes a kernel functor, then $(\mathcal{T}_\kappa, \mathcal{F}_\kappa)$ stands for the corresponding hereditary torsion theory. An object E of $\underline{\mathcal{C}}$ is said to be κ**-injective** if every exact diagram in $\underline{\mathcal{C}}$ with $C'' \in \mathcal{T}_\kappa$, may be completed by a morphism $g : C \to E$, such that $gi = f$.

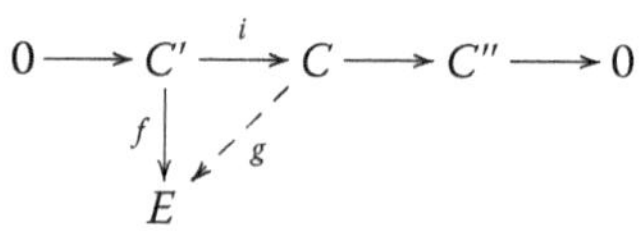

If g as above is unique as such, then E is said to be **faithfully** κ**-injective**.

Proposition 2.15
The following statements are equivalent:

1. *E is κ-injective and κ-torsion free.*

2. *E is faithfully κ-injective.*

Proof
Consider the following exact diagram in $\underline{\mathcal{C}}$:

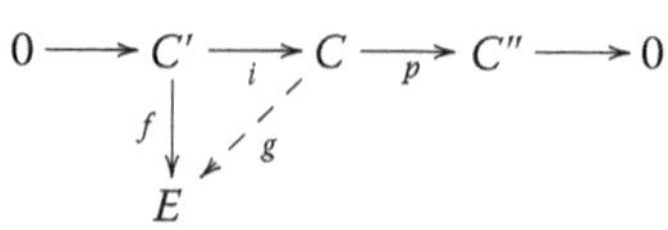

with $c'' \in \mathcal{T}_\kappa$.

Since E is κ-injective at least one morphism $g : C \to E$, such that $gi = f$, must exist. Suppose g_1, g_2 both have that property, then $(g_1 - g_2)i = 0$; hence $g_1 - g_2$ factorizes through C''; that is, there is a morphism $h : C'' \to E$ such that $g_1 - g_2 = hp$. Now $C'' \in \mathcal{T}_\kappa$, $R \in \mathcal{F}_\kappa$ yields $h = 0$ or $g_1 = g_2$. This establishes the implication 1. $\Rightarrow$ 2. Conversely, consider the diagram in $\underline{C}$:

$$\begin{array}{ccccccc} 0 & \longrightarrow & \kappa(E) & \longrightarrow & \kappa(E) & \longrightarrow & 0 \\ & \searrow & \downarrow & & & & \\ & & E & & & & \end{array}$$

Since $\kappa(E) \in \mathcal{T}_\kappa$, there is a unique extension of the zero map $0 \to E$ to $\kappa(E)$, which therefore has to be the zero map too! However, since $\kappa(E) \hookrightarrow E$ is a monomorphism it follows that $\kappa(E) = 0$. □

Proposition 2.16
Look at the exact sequence in $\underline{C}$*:*

$$0 \longrightarrow E' \xrightarrow{i} E \xrightarrow{p} E'' \longrightarrow 0$$

where E is κ-injective and E'' is κ-torsion free, and E' is κ-torsionfree. Then E' is κ-injective too.

Proof
Consider the following diagram for a given morphism $f' : C' \to E'$, where the rows in the diagram are exact:

$$\begin{array}{ccccccccc} 0 & \longrightarrow & E' & \xrightarrow{i} & E & \xrightarrow{p} & E'' & \longrightarrow & 0 \\ & & \uparrow{\scriptstyle f'} & & \uparrow{\scriptstyle f} & & \uparrow{\scriptstyle f''} & & \\ 0 & \longrightarrow & C' & \xrightarrow[j]{} & C & \longrightarrow & C'' & \longrightarrow & 0 \end{array}$$

where $C'' \in \mathcal{T}_\kappa$. Note that f is obtained from the κ-injectivity of E and f'' is just the induced quotient map.

Since $E'' \in \mathcal{F}_\kappa$ and $C'' \in \mathcal{T}_\kappa$ it follows that $f'' = 0$ of f'' factorizes through E' and $f = if_1$ for some $f_1 : C \to E'$. One easily checks that $f_1 j = f'$ and it follows that E' is κ-injective. □

Corollary 2.10
Let $0 \longrightarrow E' \xrightarrow{i} E \xrightarrow{p} E'' \longrightarrow 0$ be exact in $\underline{C}$ and assume that E' is κ-injective, $E'' \in \mathcal{T}_\kappa$ and $E \in \mathcal{F}_\kappa$; then E is isomorphic to E'.

Proof
The conditions imply that i is an essential morphism; that is, if X is a subobject of E that is nonzero then $X \cap E'$ is nonzero, as is easily seen (exercise). The assumption that

$E'' \in \mathcal{T}_\kappa$ allows us to complete the following diagram in $\underline{\mathcal{C}}$ by a morphism $j : E \to E'$ such that $ji = 1_{E'}$.

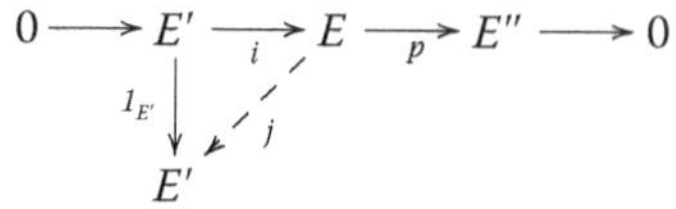

From the foregoing it follows that j is a monomorphism and then $ji = 1_{E'}$ entails that $E \cong E'$. □

The class of all faithfully κ-injective objects in $\underline{\mathcal{C}}$ is a full subcategory of $\underline{\mathcal{C}}$, which will be denoted by $\underline{\mathcal{C}}(\kappa)$ and called the **quotient category** of $\underline{\mathcal{C}}$ with respect to κ. The canonical inclusion is denoted by $i_\kappa : \underline{\mathcal{C}}(\kappa) \to \underline{\mathcal{C}}$. For C in $\mathcal{F}_\kappa$ the κ-**injective hull** of C is defined to be an essential extension $C \to E$ such that E is κ-injective and $E/C \in \mathcal{T}_\kappa$. It is clear that any κ-injective hull is in $\underline{\mathcal{C}}(\kappa)$.

Proposition 2.17
Every $C \in \mathcal{F}_\kappa$ has an essentially unique κ-injective hull.

Proof
The object C of $\underline{\mathcal{C}}$ has an injective hull E in $\underline{\mathcal{C}}$; since $C \in \mathcal{F}_\kappa$ it is clear that $E \in \mathcal{F}_\kappa$ too. Consider the exact sequence $0 \longrightarrow C \longrightarrow E \longrightarrow E/C \longrightarrow 0$ in $\underline{\mathcal{C}}$, and define $E' = E \times_{E/C} \kappa(E/C)$, which may be viewed as a subobject of E by the classical pull-back properties in Grothendieck categories. Hence $E' \in \mathcal{F}_\kappa$ and $E/E' \cong (E/C)/\kappa(E/C)$; hence $\kappa(E/E') = 0$. Apply Proposition 2.16 to conclude E' is κ-injective. On the other hand, $E'/C \cong \kappa(E/C)$ or E'/C is κ-torsion. Then let us assume that E_1', E_2' are κ-injective hulls of C. It follows that E_2' is isomorphic to a subobject E_2'' of E_1' containing C as a subobject. Because E_1' is in $\mathcal{F}_\kappa$ and also an essential extension of E_2'' that itself is faithfully κ-injective, we apply Corollary 2.10 and arrive at $E_1' \cong E_2'' \cong E_2'$. □

The κ-injective hull of $C \in \underline{\mathcal{C}}$ is denoted by $E_\kappa(C)$. Recall that in a Grothendieck category $\underline{\mathcal{C}}$ the following are equivalent for any endofunctor F:

a. F has a right adjoint.

b. F is right exact and commutes with coproducts.

Recall also that right adjoints preserve projective (inverse) limits while left adjoints preserve inductive (direct) limits.

Theorem 2.4
With notations as before: $i_\kappa : \underline{\mathcal{C}}(\kappa) \to \underline{\mathcal{C}}$ has a left adjoint.

Proof
For $C \in \underline{\mathcal{C}}$, define $a_\kappa(C) = E_\kappa(C/\kappa(C))$. This yields a functor $a_\kappa : \underline{\mathcal{C}} \to \underline{\mathcal{C}}(\kappa)$. If $f : C \to i_\kappa(D)$ is an arbitrary morphism, with $C \in \underline{\mathcal{C}}$, $D \in \underline{\mathcal{C}}(\kappa)$, then f extends to a

morphism $f_1 : C/\kappa(C) \to i_\kappa(D)$, since $i_\kappa(D) \in \mathcal{F}_\kappa$. Now $a_\kappa(i_\kappa(D))$ is faithfully κ-injective and $a_\kappa(C)/(C/\kappa(C)) \in \mathcal{T}_\kappa$, hence f_1 extends to $f' : a_\kappa(C) \to a_\kappa i_\kappa(D) = D$. Finally it is easily verified that we obtain the following isomorphism:

$$\mathrm{Hom}_{\underline{\mathcal{C}}}(\mathrm{C}, \mathrm{i}_\kappa(\mathrm{D})) \cong \mathrm{Hom}_{\underline{\mathcal{C}}(\kappa)}(\mathrm{a}_\kappa(\mathrm{C}), \mathrm{D}). \qquad \square$$

We shall write $Q_\kappa = i_\kappa a_\kappa$. For $c \in \underline{\mathcal{C}}$, the object $Q_\kappa(c)$ together with the canonical morphism $j_\kappa : C \to Q_\kappa(C)$ is called the $\underline{\mathcal{C}}$- **object of quotients of** C with respect to κ.

Proposition 2.18
Q_κ is a left exact endofunctor in $\underline{\mathcal{C}}$.

Proof
If $0 \to C' \xrightarrow{j} C$ is exact in $\underline{\mathcal{C}}$, then so is the sequence $0 \to C'/\kappa(C') \to C/\kappa(C)$. Since $Q_\kappa(C')$, $Q_\kappa(C)$ are essential extensions of $C'/\kappa(C')$, respectively $C/\kappa(C)$, it follows that $Q_\kappa(i)$ is a monomorphism. First let $C \in \mathcal{F}_\kappa$ and consider the following commutative diagram with exact top row:

$$\begin{array}{ccccccc}
0 & \longrightarrow & C' & \xrightarrow{f} & C & \xrightarrow{g} & C'' \\
 & & \downarrow & & \downarrow & & \downarrow \\
0 & \longrightarrow & Q_\kappa(C') & \xrightarrow[Q_\kappa(f)]{} & Q_\kappa(C) & \xrightarrow[Q_\kappa(g)]{} & Q_\kappa(C'')
\end{array}$$

Here $Q_\kappa(f)$ is a monomorphism and $Q_\kappa(g)Q_\kappa(f) = Q_\kappa(gf) = 0$; hence $Q_\kappa(C')$ is a subobject of $\mathrm{Ker}Q_\kappa(g)$. Then consider the exact sequence:

$$0 \longrightarrow \mathrm{Ker}Q_\kappa(g) \longrightarrow Q_\kappa(C) \longrightarrow \mathrm{Im}Q_\kappa(g) \longrightarrow 0$$

Since $Q_\kappa(C)$ is κ-injective and $\mathrm{Im}Q_\kappa(g) \in \mathcal{F}_\kappa$, we conclude that $\mathrm{Ker}Q_\kappa(g)$ is κ-injective; hence faithfully κ-injective. Moreover $\mathrm{Ker}Q_\kappa(g)/C' \cong Q_\kappa(C)/C$ is in $\mathcal{T}_\kappa$; hence Corollary 2.10 yields $Q_\kappa(C') = \mathrm{Ker}Q_\kappa(g)$. In general, that is, when C is not necessarily in $\mathcal{F}_\kappa$ we consider

$$0 \longrightarrow C' \xrightarrow{f} C \longrightarrow C'' \longrightarrow 0$$

and define $D = C \times_{C''} \kappa(C'')$, the pre-image of $\kappa(C'')$ in C.

Then $\kappa(C)$ is clearly a subobject of D in $\underline{\mathcal{C}}$. Also $\mathrm{Im}f$ is a subobject of D and $D/\mathrm{Imf} \cong \kappa(C'')$. Therefore $D/\kappa(C)$ contains $(\mathrm{Im}f + \kappa(C))/\kappa(C)$ such that modulo the latter it is κ-torsion. We obtain an exact sequence:

$$0 \longrightarrow D/\kappa(C) \longrightarrow C/\kappa(C) \longrightarrow C''/\kappa(C'') \longrightarrow 0$$

where $\kappa(C/\kappa(C)) = 0$. Now we have reduced the problem to the torsion-free case because we obtain an exact sequence

$$0 \longrightarrow Q_\kappa(D/C) \longrightarrow Q_\kappa(C) \longrightarrow Q_\kappa(C'')$$

where $Q_\kappa(D/C) = Q_\kappa(\mathrm{Im}f + \kappa(C)/\kappa(C)) = \mathrm{Im}Q_\kappa(f)$. $\square$

Note that a_κ has right adjoint i_κ but Q_κ need not have a right adjoint, in fact Q_κ need not even be a right exact functor. For further application of these techniques to categories of sheaves or presheaves it is worthwhile to present some basic facts about Giraud subcategories of (complete) Grothendieck categories. This allows us to apply the "reflector" approach to localization theory; the presentation here is close to Section 2.3 in F. Van Oystaeyen, A. Verschoren, *Reflectors and Localization. Application to Sheaf Theory*, Lect. Notes in Pure and Applied Mathematics Vol. 41, M. Dekker, New York, 1978 [48]. This approach also allows a general treatment of compatibility of kernel functors and commuting properties of localization functors, that is, exactly the topic recognized in noncommutative topology with respect to the relations between noncommutative space and the commutative shadow. Compatibility of localization goes back to F. Van Oystaeyen, "Compatibility of Kernel Functors and Localization Functors" [45].

The consideration of Giraud subcategories of (complete) Grothendieck categories prepares for the study of **sheaves** as a subcategory of **presheaves**. So we look at a complete Grothendieck category $\underline{\mathcal{P}}$; a full subcategory $\underline{\mathcal{S}}$ of $\underline{\mathcal{P}}$ is called **reflective** if the inclusion functor $i : \underline{\mathcal{S}} \to \underline{\mathcal{P}}$ has a left adjoint a, called the reflector of $\underline{\mathcal{S}}$ in $\underline{\mathcal{P}}$; that is, for $P \in \underline{\mathcal{P}}$, $S \in \underline{\mathcal{S}}$ there is a natural isomorphism $\mathrm{Hom}_{\underline{\mathcal{P}}}(\mathrm{P}, \mathrm{iS}) \cong \mathrm{Hom}_{\underline{\mathcal{S}}}(\mathrm{aP}, \mathrm{S})$ with canonical natural transforms $p : ai \to 1_{\underline{\mathcal{S}}}$ and $q : 1_{\underline{\mathcal{P}}} \to ia$. The couple $(aP, q_P : P \to iaP)$ has the following universal property: every $\underline{\mathcal{P}}$-morphism $f : P \to iS$ with $S \in \underline{\mathcal{S}}$ factorizes in a unique way as follows:

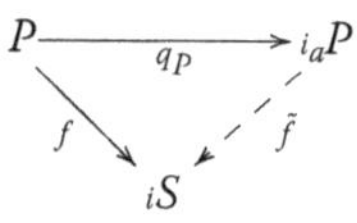

The morphism q_P is called the **reflection of** P. □

Proposition 2.19
Let $\underline{\mathcal{S}}$ be a reflective subcategory of a (complete) Grothendieck category $\underline{\mathcal{P}}$; then $\underline{\mathcal{S}}$ is complete and co-complete. If the reflector of $\underline{\mathcal{S}}$ in $\underline{\mathcal{P}}$ is left exact, then $\underline{\mathcal{S}}$ has exact direct limits and a generator.

A subcategory of $\underline{\mathcal{P}}$ with a left exact reflector is called a ***Giraud subcategory*** *of $\underline{\mathcal{P}}$. From the definition it follows that a Giraud subcategory of a (complete) Grothendieck category is itself a (complete) Grothendieck category and the reflector is exact, whereas the inclusion functor $\underline{\mathcal{S}} \to \underline{\mathcal{P}}$ is in general only left exact.*

Let $\mathcal{T}$ be the class of objects C in $\underline{\mathcal{P}}$ for which $a(C) = 0$ and let $\mathcal{F}$ consist of subobjects in $\underline{\mathcal{P}}$ of objects of $\underline{\mathcal{S}}$.

Proposition 2.20
$P \in \underline{\mathcal{P}}$ is in $\mathcal{F}$ exactly when $q_P : P \to ia\mathcal{P}$ is a monomorphism.

Proof
If q_P is a monomorphism, then P is in $\mathcal{F}$ since $a(P) \in \mathcal{S}$. Conversely, consider $P \in \mathcal{F}$ and $0 \to P \to i(S)$ with $S \in \underline{\mathcal{S}}$. Commutativity of the following diagram in $\underline{\mathcal{P}}$:

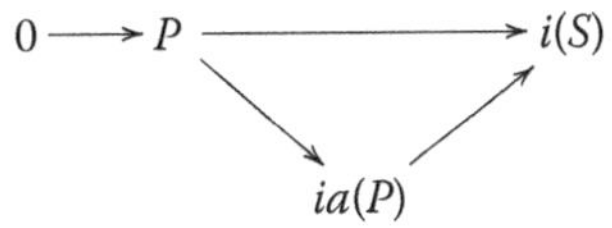

leads to the conclusion that q_p is a monomorphism.

$\mathcal{F}$ may be viewed as a full complete subcategory of $\underline{\mathcal{P}}$ easily verified to be a reflective subcategory of $\underline{\mathcal{P}}$ with epimorphism reflector aj where a is the reflector of $\underline{\mathcal{S}}$ and $j : \mathcal{F} \to \underline{\mathcal{P}}$ the canonical inclusion function. The objects of $\mathcal{F}$ are said to be separated. □

Proposition 2.21
If $P \in \underline{\mathcal{P}}$ is separated, then $ia(P)$ is an essential extension of P in $\underline{\mathcal{P}}$.

Proof
Let Q be a subobject of $ia(P)$ in $\underline{\mathcal{P}}$ and assume that $Q \times_{ia(P)} P = 0$. Then $0 = a(Q \times_{ia(P)} P) = a(Q) \times_{a(P)} a(P) \cong a(Q)$. Since Q is a subobject of a separated object, it is itself separated; that is, the canonical $Q \to ia(Q)$ is a monomorphism and thus $Q = 0$ follows from $0 = a(Q)$. □

Proposition 2.22
Consider an exact sequence in $\underline{\mathcal{P}}$:

$$0 \longrightarrow P' \longrightarrow \mathcal{P} \longrightarrow P'' \longrightarrow 0$$

1. *If $P' \in \underline{\mathcal{S}}$, $P \in \mathcal{F}$, then $P'' \in \mathcal{F}$.*
2. *If $P'' \in \mathcal{F}$, $P \in \underline{\mathcal{S}}$, then $P' \in \underline{\mathcal{S}}$.*

Proof
Instead of providing a direct proof we can derive it directly from the next proposition, which reduces it to the torsion theory situation and Proposition 2.16 as well as Corollary 2.10 (with a slight rephrasing). □

Proposition 2.23
The couple $(\mathcal{T}, \mathcal{F})$ determines a torsion theory in $\underline{\mathcal{P}}$ such that its quotient category is equivalent to $\underline{\mathcal{S}}$.

Proof

Since a is exact, $\mathcal{T}$ is closed under subobjects, quotient objects, and extensions. Since a has a right adjoint, it preserves coproducts, hence $\mathcal{T}$ is a torsion class.

Obviously for $T \in \mathcal{T}$, $S \in \underline{S}$ we have $\mathrm{Hom}_{\underline{P}}(\mathrm{T}, \mathrm{i(S)}) = 0$ and also $\mathrm{Hom}_{\underline{P}}(\mathrm{T}, \mathrm{F}) = 0$ for every F in $\mathcal{F}$. Conversely, if $\mathrm{Hom}_{\underline{P}}(\mathrm{T}, \mathrm{P}) = 0$ for all $T \in \mathcal{T}$, then from $\mathrm{Ker} q_P \in \mathcal{T}$ we obtain that $\mathrm{Ker} q_p = 0$, hence $P \in \mathcal{F}$. Consequently $\mathcal{F}$ may be considered as the torsion-free class corresponding to $\mathcal{T}$. □

The kernel functor associated to $(\mathcal{T}, \mathcal{F})$ will be denoted by α. Clearly if $P \in \underline{P}$ then:

$$\alpha(P) = \sum\{P', 0 \longrightarrow P' \longrightarrow P \quad \text{and} \quad a(P') = P'\}$$

A Giraud subcategory of $\underline{P}$ is said to be **strict** if it is closed under $\underline{P}$-isomorphisms.

Observation 2.5

If κ is a kernel functor for a (complete) Grothendieck category $\underline{P}$, then $\underline{P}(\kappa)$ is a strict Giraud subcategory of $\underline{P}$.

Proof

From Proposition 2.17.

From what we have already learned it follows easily that there is a bijective correspondence between torsion theories for $\underline{P}$ and strict Giraud subcategories of $\underline{P}$; this is otherwise known as **Gabriel's Theorem**.

At times we have used the term (complete) Grothendieck category; in fact this indicates that the original statement of the result used the completeness as an extra assumption. It was only after the proof of the Gabriel-Popescu, theorem (first by Popescu, but with a gap solved by Gabriel) that it followed that a Grothendieck category with a generator has enough injective objects and also that every Grothendieck category is complete. □

Theorem 2.5: Gabriel-Popescu

Let $\underline{C}$ be any Grothendieck category with generator G; put $R = \mathrm{Hom}_{\underline{C}}(\mathrm{G}, \mathrm{G})$ and let $M : \underline{C} \to R$-mod be the functor $C \to \mathrm{Hom}_{\underline{C}}(\mathrm{G}, \mathrm{C}) = \mathrm{M(C)}$.

1. *M is full and faithful.*
2. *M induces an equivalence between $\underline{C}$ and $(R - \mathrm{mod})(\kappa)$, where κ is the largest kernel functor in R-mod for which all modules $M(C)$, $C \in \underline{C}$, are faithfully κ-injective.*

Corollary 2.10

Every object C in $\underline{C}$ has an extension that is an injective object of $\underline{C}$. Every $\underline{C}$ is complete!

The categorical approach to localization theory has an undeniable elegance, but now we have a less obvious notion of the noncommutative composition at hand, unless we start to compare localization of a strict Giraud category $\underline{S}$ to localization of $\underline{P}$.

Proposition 2.24
An object E of $\underline{\mathcal{S}}$ is injective in $\mathcal{S}$ if and only if $i(E)$ is injective in $\underline{\mathcal{P}}$.

Proof
If $i(E)$ is injective in $\underline{\mathcal{P}}$, then E is injective in $\underline{\mathcal{S}}$ because i is left exact. Conversely, suppose that E is injective in $\underline{\mathcal{S}}$ and consider a diagram in $\underline{\mathcal{S}}$:

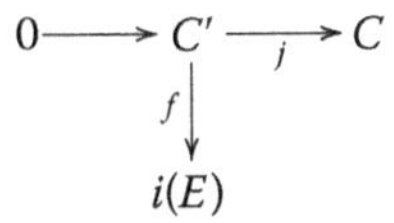

yielding a commutative diagram in $\underline{\mathcal{P}}$:

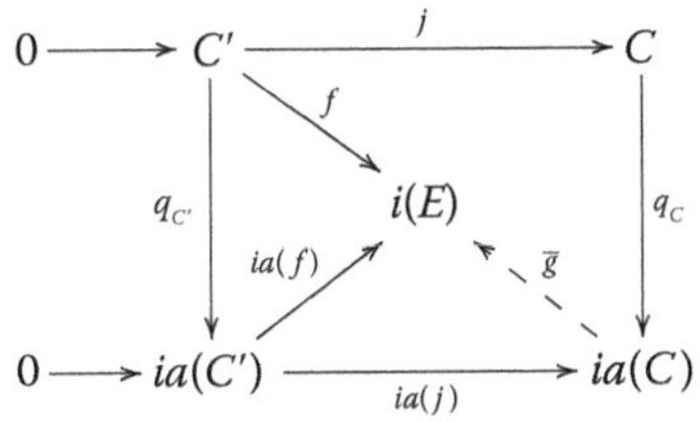

where existence of $\overline{g}$ follows from the injectivity of E in $\underline{\mathcal{S}}$: $ia(f) = \overline{g} \circ ia(j)$. Put $g = \overline{g}q_C$; then we find: $gj = \overline{g}q_C j = \overline{g}ia(j)q_{C'} = ia(f)q_{C'} = f$, finishing the proof. □

For $C \in \underline{\mathcal{S}}$, respectively $C \in \underline{\mathcal{P}}$, the injective hull of C in $\underline{\mathcal{S}}$, respectively in $\underline{\mathcal{P}}$, will be denoted by $E^s(C)$, respectively $E^p(C)$.

Lemma 2.22

1. *For $S \in \underline{\mathcal{S}}$, $i(E^s(S))$ is an essential extension of $i(S)$ in $\underline{\mathcal{P}}$.*
2. *Let $S \in \underline{\mathcal{S}}$; then $E^p(i(S)) = iE^s(S)$; that is, the hull in $\underline{\mathcal{P}}$ of an object in $\underline{\mathcal{S}}$ is in $\underline{\mathcal{S}}$ too.*

Proof

1. Let P be a subobject of $i(E^s(S))$ in $\underline{\mathcal{P}}$ such that $P \times_{iE^s(S)} i(S) = 0$. Exactness of a yields $0 = a(0) = a(P \times_{iE^s(S)} i(S)) = a(P) \times_{ai(E^s(S))} ai(S) = a(P) \times_{E^s(S)} S$, contradicting the fact that $E^s(S)$ is an essential extension of S in $\underline{\mathcal{S}}$ since $a(P) \in \underline{\mathcal{S}}$.
2. The foregoing implies that both $i(E^s(S))$ and $E^p(i(S))$ are essential extensions of $i(S)$ in $\underline{\mathcal{P}}$; thus we arrive at a commutative diagram in $\underline{\mathcal{P}}$:

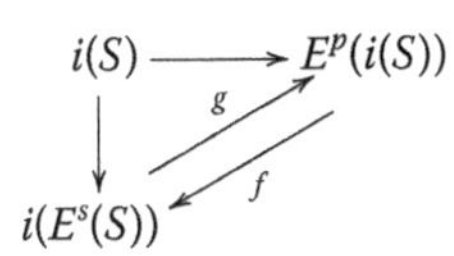

where f exists by definition of E^p, and it is a monomorphism, moreover g is a monomorphism too. Since $E^p(i(S))$ is a maximal essential extension of $i(S)$ in $\underline{P}$, it follows from

$$0 \longrightarrow E^p(i(S)) \underset{f}{\longrightarrow} iE^s(S)$$

that $E^p(i(S)) \cong i(E^s(S))$. □

Corollary 2.11

If $P \in \underline{P}$ is separated, then $E^p(P)$ is separated. Indeed we have a commutative diagram of monomorphisms in $\underline{P}$:

$$\begin{array}{ccc} P & \longrightarrow & ia(P) \\ \downarrow & & \downarrow \\ E^p(P) & \longrightarrow & E^p(ia(P)) \end{array}$$

Since $E^p(ia(P)) \cong iE^s(a(P))$, it follows that $E^p(P)$, being a subobject of $E^s(a(P))$ in $\underline{P}$, is separated.

Proposition 2.25

If $P \in \underline{P}$ is separated, then we have:

$$ia(E^p(P)) \underset{(1)}{=} E^p(ia(P)) \underset{(2)}{=} i(E^s(a(P))) \underset{(3)}{=} E^p(P)$$

Proof

1. We obviously have the following monomorphisms:

   $$P \longrightarrow ia(P), \qquad E^p(P) \longrightarrow E^p(ia(P)), \qquad iaE^p(P) \longrightarrow E^p(ia(P))$$

   Now $E^p(ia(P))$ is essential over $ia(P)$ and this in turn is essential over P in $\underline{P}$; that is, $E^p(ia(P))$ is essential over P. The equality (1) follows if we establish that $ia(E^P(P))$ is injective in $\underline{P}$; that is $aE(P)$ is injective in $\underline{S}$ (see 2.24.). Consider an exact sequence $0 \longrightarrow S' \underset{s}{\longrightarrow} S$ and a given $f : S' \longrightarrow aE^p(P)$in $\underline{S}$. The pull-back properties yield:

   $$\begin{array}{ccccc} S'_1 = iS' \underset{iaE^p(P)}{\times} E^p(P) & \xrightarrow{s'} & iS' & \xrightarrow{is} & iS \\ {\scriptstyle f_1}\downarrow & & {\scriptstyle if}\downarrow & \swarrow {\scriptstyle ig} & \\ E^p(P) & \longrightarrow & iaE^p(P) & & \end{array}$$

   where s' is a monomorphism, thus $(is)s'$ is monomorphic. Obviously:

   $$a(S_1') = S' \underset{aE^p(P)}{\times} aE^p(P) = S'$$

By the injectivity of $E^p(P)$ in $\underline{\mathcal{P}}$, there is a $\underline{\mathcal{P}}$-morphism $g_1 : iS \to E^p(P)$, such that $g(is)s' = f_1$. Put $g = a(g_1) : S \to aE^p(P)$. Let j be the isomorphism $aS_1' \to S'$. Then we have $af_1 = fj = q(g_1)sa(s') = gsj$, with $fj = gsj$, hence $f = gs$ since j is an isomorphism and thus in particular an epimorphism in $\underline{\mathcal{S}}$.

2. The equality (2) is a direct consequence of Lemma 2.2.2(2).
3. $E^p(P)$ is separated because P is separated (Corollary 2.11). Then $ia(E^p(P))$ is essential over $E^p(P)$, hence over P in $\underline{\mathcal{P}}$. Therefore $E^p(P) = iaE^p(P)$ because $E^p(P)$ is a maximal essential extension of P in $\underline{\mathcal{P}}$. □

Consider kernel functors κ, κ' in $\underline{\mathcal{P}}$. Then $\kappa \geq \kappa'$ if and only if $\mathcal{T}_\kappa \supset \mathcal{T}_{\kappa'}$ or equivalently $\kappa(P) \supset \kappa'(P)$ for every object P of $\underline{\mathcal{P}}$. For kernel functors κ and κ' in $\underline{\mathcal{P}}$ we say that κ' is Q_κ-compatible if $\kappa' Q_\kappa = Q_\kappa \kappa'$.

Lemma 2.23
Suppose κ' is Q_κ-compatible.

1. *If P is in $\mathcal{F}_{\kappa'}$, then $Q_\kappa(P)$ is in $\mathcal{F}_{\kappa'}$; the converse holds when P is κ-torsion free.*
2. *If P is in $\mathcal{T}_{\kappa'}$, then $Q_\kappa(P)$ is in $\mathcal{T}_{\kappa'}$; the converse holds in case $P \in \mathcal{F}_\kappa$.*

Proof
Exercise. □

To any strict Giraud subcategory $\underline{\mathcal{S}}$ of $\underline{\mathcal{P}}$ we have associated a kernel functor α (see remark Proposition 2.23).

The α- **compatible kernel functors** (Q_α-compatible) are sometimes called **$\underline{\mathcal{S}}$-compatible** kernel functors. Hence, κ in $\underline{\mathcal{P}}$ is $\underline{\mathcal{S}}$-compatible exactly when $ia\kappa = \kappa ia$. If $ia\kappa ia = \kappa ia$, that is, κ takes objects of $\underline{\mathcal{S}}$ to objects of $\underline{\mathcal{S}}$, then κ is said to be **inner** in $\underline{\mathcal{S}}$. If κ is inner in $\underline{\mathcal{S}}$, then the functor κ is denoted by κ^S; in general κ^S need not be a kernel functor in $\underline{\mathcal{S}}$.

Proposition 2.26
Let κ be $\underline{\mathcal{S}}$-compatible in $\underline{\mathcal{P}}$; then κ^s is a kernel functor in $\underline{\mathcal{S}}$.

Proof
For S in $\underline{\mathcal{S}}$, $\kappa(iS) = \kappa(iai(S))$; hence κ^S is inner in $\underline{\mathcal{S}}$. Therefore $a\kappa(i(S)) = \kappa^s(S)$ and it is clear that κ^s is a left exact subfunctor of the identity in $\underline{\mathcal{S}}$. Furthermore we easily calculate:

$$\kappa^s(S/\kappa^s(S)) = a\kappa(ia(iS/I\kappa^s(S)))$$
$$= a\kappa(iS/i\kappa^s(S)) = a\kappa(iS/\kappa(iS)) = a(0) = 0$$

(*Note:* we simplified notation by dropping some brackets in the notation). □

Earlier we mentioned the gen-topology induced on a lattice by taking the intervals $[0, \lambda]$, $\lambda \in \Lambda$; now comparing the lattice of kernel functors on $\underline{\mathcal{P}}$ to what the Zariski topology would be if $\underline{\mathcal{P}}$ were C-mod for some commutative ring, we know we have to look at the opposite lattice, and therefore the sets $[\kappa, 1]$ get our attention. These are well behaved in terms of compatibility condition because of the following result.

Proposition 2.27
Look at kernel functors κ and κ', for $\underline{\mathcal{P}}$ such that $\kappa \geq \kappa'$; then κ is $Q(\kappa')$-compatible.

Proof
Let $q_{\kappa(P)} : \kappa(P) \to Q_{\kappa'}(\kappa(P))$ be the reflection of $P \in \underline{\mathcal{P}}$ in $\underline{\mathcal{P}}(\kappa')$. The following sequence is exact:

$$0 \to \mathrm{Im} q_{\kappa(P)} \to Q_{\kappa'}(\kappa(P)) \to \mathrm{Coim} q_{\kappa(P)} \to 0$$

Clearly, $\mathrm{Im} q_{\kappa(P)}$ is κ-torsion; moreover $\kappa \geq \kappa'$ implies that $\mathrm{Coim} q_{\kappa(P)}$ is κ-torsion. Hence, $Q_{\kappa'}(\kappa(P))$ is a subobject $\kappa Q_{\kappa'}(P)$. Conversely, since $\mathcal{F}_\kappa \subset \mathcal{F}_{\kappa'}$ we have that $P/\kappa(P) \in \mathcal{F}_{\kappa'}$ so we have a monomorphism: $0 \to P/\kappa(P) \to Q_{\kappa'}(P/\kappa(P))$. Then $P/\kappa(P) \cap \kappa Q_{\kappa'}(P/\kappa(P)) = 0$ implies that $Q_{\kappa'}(P/\kappa(P)) = 0$. Finally, by the exactness of

$$0 \to Q_{\kappa'}(\kappa(P)) \to Q_{\kappa'}(P) \to Q_{\kappa'}(P/\kappa(P))$$

we obtain that $\kappa Q_{\kappa'}(P) \subset Q_{\kappa'}(\kappa(P))$. □

Theorem 2.6
For a strict Giraud subcategory $\underline{\mathcal{S}}$ of $\underline{\mathcal{P}}$ and an $\underline{\mathcal{S}}$-compatible kernel functor κ for $\underline{\mathcal{P}}$ we have $S \in \underline{\mathcal{S}}$ is (faithfully) κ^s-injective if and only if $i(S)$ is (faithfully) κ-injective.

Proof
Assume that $i(S)$ is κ-injective. Consider a diagram in $\underline{\mathcal{S}}$:

$$\begin{array}{ccccccccc} 0 & \longrightarrow & S_1 & \longrightarrow & S_2 & \longrightarrow & S_2/S_1 & \longrightarrow & 0 \\ & & \big\downarrow f & & & & & & \\ & & S & & & & & & \end{array}$$

where $\kappa^s(S_2/S_1) = S_2/S_1$.
In $\underline{\mathcal{P}}$ we obtain a diagram, applying i to the above:

$$\begin{array}{ccccccc} 0 & \longrightarrow & i(S_1) & \longrightarrow & i(S_2) & \longrightarrow & i(S_2)/i(S_1) \\ & & \big\downarrow if & & & & \\ & & i(S) & & & & \end{array}$$

where $i(S_2)/i(S_1)$ is subobject of $i(S_2/S_1)$.
Since S_2/S_1 is κ^s-torsion and $S_2/S_1 = a(iS_2/iS_1)$, it follows Lemma 2.23(2) that iS_2/iS_1 is κ-torsion, so there exists a $g' : iS_2 \to iS$ completing the above diagram. Then it is clear that $a(g')$ completes the diagram in $\underline{\mathcal{S}}$ from which we started.

Conversely, let S be κ^s-injective and consider an exact sequence in $\underline{P}$:

$$\begin{array}{ccccccc} 0 & \longrightarrow & P_1 & \longrightarrow & P_2 & \longrightarrow & P_2/P_1 \\ & & \downarrow{\scriptstyle f} & & & & \\ & & i(S) & & & & \end{array}$$

where $P_2/P_1 \in \mathcal{T}_\kappa$. Since a is exact we obtain the following diagram in $\underline{S}$:

$$\begin{array}{ccccccc} 0 & \longrightarrow & aP_1 & \longrightarrow & aP_2 & \longrightarrow & a(P_2/P_1) \\ & & \downarrow{\scriptstyle af} & & & & \\ & & S & & & & \end{array}$$

Again, by Lemma 2.23, it follows that $a(P_2/P_1)$ is κ^s-torsion; therefore, there exists an $\underline{S}$-morphism $g' : aP_2 \to S$ completing the diagram. Let g be the map $(ig')q_{P_2} : P_2 \to i(S)$, and it is easily verified that g extends f as desired. Since $i(S)$ is κ-torsion free if and only if S is κ^s-torsion free (because $i(S)$ is separated and Lemma 2.23) we may apply Proposition 2.15. □

Proposition 2.28
With notation as before: Let κ be an $\underline{S}$-compatible kerel functor for $\underline{P}$ and consider an object S of $\underline{S}$ that is κ^s-torsion free, $i(E_{\kappa^s}(S)) \cong E_\kappa(i(S))$.

Proof
Lemma 2.23 yields: $i(S) \in \mathcal{F}_\kappa$. The foregoing and the fact that $E_{\kappa^s}(S)$ is faithfully κ^s-injective imply that $iE_{\kappa^s}(S)$ is faithfully κ-injective. Furthermore, $iE_{\kappa^s}(S)/iS$ is κ-torsion in $\underline{P}$ because $E_{\kappa^s}(S)/S$ is κ^s-torsion in $\underline{S}$. But $E_\kappa(iS)$ is unique up to isomorphism in $\underline{P}$ with the properties mentioned above; therefore, we arrive at $E_\kappa(iS) \cong iE_{\kappa^s}(S)$. □

Corollary 2.12
If κ is an $\underline{S}$-compatible kernel functor for $\underline{P}$ and $S \in \underline{S}$, then $iQ_{\kappa^s}(S) = Q_\kappa(iS)$.

Proof
If S is κ^s-torsion free, then the statement follows from the foregoing proposition. In general:

$$iQ_\kappa(S) = iE_{\kappa^s}(S/\kappa^s(S)) = E_\kappa(ia(iS/\kappa(iS)))$$

Of course, $iS/\kappa(iS)$ is separated; thus $ia(iS/\kappa(iS))$ is an essential extension in $\underline{P}$ of $iS/\kappa(iS)$. Consequently $i(S/\kappa^s(S))$ is κ-torsion free and

$$Q_\kappa(iS) = E_\kappa(iS/\kappa(iS)) \cong E_\kappa(ia(iS/\kappa(iS))).$$ □

Proposition 2.29
Let $\underline{S}$ be a strict Giraud subcategory of $\underline{P}$ and let κ be a kernel functor in $\underline{P}$ such that $\kappa \geq \alpha$ where α corresponds to $\underline{S}$. Then we have: $Q_\kappa(P) \in i\underline{S}$ for every $P \in \underline{P}$.

Proof
Put $\overline{P} = P/\kappa(P)$. Since $\kappa \geq \alpha, \alpha(\overline{P}) = 0$; hence $\overline{P}$ is separated with respect to $\underline{S}$. From $a(ia(\overline{P})/\overline{P}) = 0$ it follows that $ia\overline{P}/\overline{P}$ is α-torsion, hence κ-torsion. Faithful κ-injectivity of $Q_\kappa(P)$ gives rise to the following commutative diagram in $\overline{\mathcal{P}}$:

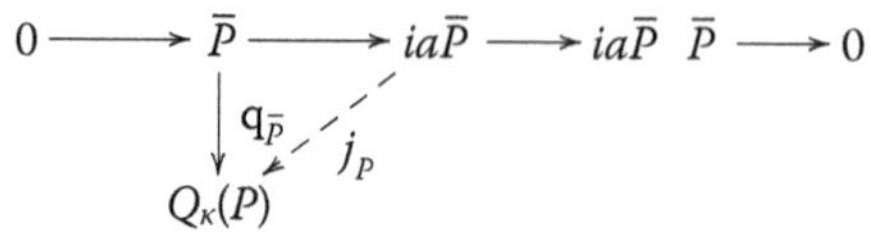

Since $q_{\overline{P}}$ is a monomorphism and since $ia\overline{P}$ is an essential extension of $\overline{P}$ in $\underline{P}$, j_P is a monomorphism and thus $Q_\kappa(ia\overline{P}) = Q_\kappa(P)$. Because of Proposition 2.27, κ is $\underline{S}$-compatible; then Corollary 2.12 applies and we obtain $iQ_{\kappa^s}(a\overline{P}) = Q_\kappa(ia\overline{P}) = Q_\kappa(P)$, and finally this yields that $Q_\kappa(P)$ is in $i\underline{S}$. □

Let us now construct some kernel functors that are $\underline{S}$-compatible.

Consider a nonzero object P in $\underline{P}$ and let $K(P)$ be the class of kernel functors κ for $\underline{P}$ such that $P \in \mathcal{F}_\kappa$. If $P' \in \underline{P}$ is essential over P in $\underline{P}$, then obviously $K(P) = K(P')$. Define κ_P for $\underline{P}$ by putting:

$$\kappa_P(Q) = \cap\{\ker g,\ g \in \mathrm{Hom}_{\underline{P}}(Q, E^P(P))\}$$

Proposition 2.30
With notation as before:

i. *κ_P is a kernel functor for $\underline{P}$ and $\kappa_P \in K(P)$.*

ii. *If κ is a kernel functor for $\underline{P}$, then $\kappa \in K(P)$ if and only if $\kappa \leq \kappa_P$.*

Proof

i. Straightforward (note that there is a monomorphism $0 \to E^P(P) \to E^P(P)$; hence $\kappa_P(E^P(P)) = 0$ and $\kappa_P \in K(P)$).

ii. If $\kappa \leq \kappa_P$, then $\kappa \in K(P)$ obviously. Conversely, if $\kappa \in K(P)$ for some kernel functor κ for $\underline{P}$, look at an arbitrary morphism $g : Q \to E^P(P)$, in $\underline{P}$. Clearly $\kappa(Q) \subset \mathrm{Ker} g$; hence $\kappa(Q) \subset \kappa_P(Q)$, for every Q in $\underline{P}$; thus $\kappa \leq \kappa_P$. □

Theorem 2.7
If $S \in \underline{S}$, then $\kappa_{i(S)}$ is an $\underline{S}$-compatible kernel functor for $\underline{P}$. Conversely, if P in $\underline{P}$ is such that κ_P is $\underline{S}$-compatible, then there is an S in $\underline{S}$ such that $\kappa_P = \kappa_{i(S)}$; moreover $S = a(P)$.

Proof
We first check that $ia\kappa_{i(S)}(P) = \kappa_{i(S)}(iaP)$, and we may assume that we have replaced S by $E^s(S)$ and $i(S)$ by $E^P(i(S))$. Then the problem reduces to proving:

$$\cap \mathrm{Ker} g,\ g \in \mathrm{Hom}_{\underline{P}}(iaP, iS) = ia(\cap\{\mathrm{Ker} g,\ g \in \mathrm{Hom}_{\underline{P}}(P, iS)\})$$

Since $\underline{S}$ is a full subcategory of $\underline{P}$ and since a is a right adjoint of i, we do obtain the following isomorphisms in $\underline{Ab}$:

$$\mathrm{Hom}_{\underline{P}}(\mathrm{iaP}, \mathrm{iS}) \cong \mathrm{Hom}_{\underline{S}}(\mathrm{aP}, \mathrm{S}) = \mathrm{Hom}_{\underline{P}}(\mathrm{P}, \mathrm{iS}).$$

Since $\underline{S}$ is a Giraud subcategory of $\underline{P}$, we have that

$$\cap\{\mathrm{Ker} g,\ g \in \mathrm{Hom}_{\underline{S}}(\mathrm{aP}, \mathrm{S})\} \text{ is an object of } \underline{S}.$$

Conversely, if $P \in \underline{P}$ is such that κ_P is $\underline{S}$-compatible, then $\kappa_P(iaP) = ia\kappa_P(P) = ia(0) = 0$, hence $\kappa_P \leq \kappa_{ia(P)}$. On the other hand:

$$\kappa_{ia(P)}(P) = \cap\{\mathrm{Ker} g,\ g \in \mathrm{Hom}_{\underline{P}}(\mathrm{P}, \mathrm{E}^{\mathrm{P}}(\mathrm{iaP}))\}.$$

In view of Lemma 2.22(2) we arrive at:

$$\begin{aligned} \kappa_{ia(P)}(P) &= \cap\{\mathrm{Ker} g,\ g \in \mathrm{Hom}_{\underline{P}}(P, iE^s(aP))\} \\ &= \cap\{\mathrm{Ker} g,\ g \in \mathrm{Hom}_{\underline{P}}(aP, E^s(aP))\} = 0 \end{aligned}$$

Finally we find $0 = \kappa_{ia(P)}(P)$; thus $\kappa_{ia(P)} \leq \kappa_P$. □

The foregoing techniques may be applied to some interesting special cases. Of course $\underline{P} = R$-mod for some noncommutative ring R is of interest, but so is the case where $\underline{P}$ is the category of presheaves over a small category X with values in a Grothendieck category $\underline{C}$. We return to this later.

2.6.1 Project: Microlocalization in a Grothendieck Category

In the algebraic geometry of associative algebras (see [46]), a particularly interesting case is presented by filtered algebras that are "almost commutative" in the sense that the associated graded ring is a commutative ring. Their noncommutative site may be viewed as being quantum-commutative in the sense that the topology defined in terms of microlocalzation functors is in fact a commutative one. Roughly speaking (see [44] for full detail) the microlocalization is obtained from a completion with respect to a localized filtration. This project is to develop such a technique for arbitrary Grothendieck categories; this can then be continued along the lines of Chapter 3, leading to canonical microtopologies. There may be a benefit of this to sheaf theory, but at this moment there are no obvious applications of this technique outside the algebraic theory already covered in [46]; however, the consideration of categories of topologized objects is natural in the context we have developed, so it is not unlikely that new applications of the microlocalizations may be discovered. Let κ be a kernel functor on the Grothendieck category $\underline{C}$ and let $\mathcal{T}_\kappa$ denote its torsion class (see Definition 2.8). To an arbitrary object M of $\underline{C}$ we may associate a filter $\mathcal{L}(\kappa, M)$ consisting of all subobjects N of M in $\underline{C}$ such that $M/N \in \mathcal{T}_\kappa$. It is clear that $\mathcal{L}(\kappa, M)$ is closed under the lattice operations $\wedge(= \cap)$ and $\vee(= \Sigma)$ defined in $\mathcal{L}(M)$, the (big) lattice of subobjects of M in $\underline{C}$ (see also the remarks following Lemma 2.3). The (big) lattice $\mathcal{L}(\kappa, M)$ need not have a 0 but we may formally add such if we wish. In any case we may view

$\mathcal{L}(\kappa, M)$ as a topologization of M and we may define $\widehat{M}_\kappa$ in $\underline{\mathcal{C}}$ as $\varprojlim_{N\in\mathcal{L}(\kappa,M)}(M/N)$. The relation between $\widehat{M}_\kappa$ and $(Q_\kappa(M))^\wedge$ is easily investigated. This provides us with a general notion of microlocalization, denoted Q^μ_κ for a torsion theory $(\mathcal{T}_\kappa, \mathcal{F}_\kappa)$. When Q_κ is an exact functor the properties of Q^μ_κ have to be investigated.

2.7 The Fundamental Examples II: *L(H)*

Let H be a complex Hilbert space and consider the set $L(H)$ of closed subspaces of H. The set $L(H)$ becomes a complete lattice if we define: for $U, V \in L(H)$, $U \wedge V = U \cap V$, $U \vee V = \overline{(U+V)}$, where $\overline{(-)}$ denotes the closure in H. In $L(H)$ we also have a **complement**, associating the orthogonal complement $U^\perp$ to U in H. It is clear that $L(H)$ satisfies axioms A.1, ..., A.9, but not A.10 (look at a space of finite codimension in H and a basis for its orthogonal complement; try to use this global cover to induce a cover on a line in the orthogonal complement disjoint from the chosen basis). Consequently it is impossible here to use Lemma 2.20 to obtain sheaves over $L(H)$! However, we shall have other techniques available that will allow the construction of sheaves (and sheafification) over the generalized Stone space, which will be introduced later in this section. Consider the algebra $\mathcal{L}(H)$ of bounded linear operators on H. Associating to U in $L(H)$ the orthogonal projection P_U onto U viewed as an element in $\mathcal{L}(H)$, then we see that the lattice $L(H)$ is isomorphic to the lattice $P(\mathcal{L}(H))$ of orthogonal projections in $\mathcal{L}(H)$ which are exactly the idempotent elements of $\mathcal{L}(H)$. It is well known (and easily verified) that $L(H)$ is not distributive due to the fact that P_U and P_V need not commute. From the theory on noncommutative topologies in Chapter 2, we expect that idempotency of $P_U P_V$ and $P_V P_U$ would lead to the commutativity of P_U and P_V. In fact this is the case, but an even stronger result holds because only one such product has to be considered (I thank my colleague Jan van Casteren for some discussions about the analytical aspects).

Proposition 2.31
With notation as introduced above: if $(P_U P_V)^2 = P_U P_V$ then P_U and P_V commute and $P_U P_V = P_{U\cap V}$.

Proof
Observe that $P_U = P_U^*$, $P_V = P_V^*$. Put $T = P_U P_V - P_V P_U$, then $T^* = -T$ and we easily calculate:

$$P_U T P_V = P_V T P_U = P_U T P_U = P_V T P_V = 0$$

For $f \in U^\perp \cap V^\perp$ we have:

$$Tf = P_U P_V f - P_V P_U f = P_U.0 - P_V.0 = 0$$

Clearly, if $f \in U$, then $Tf \in U + V$ but also $Tf \in U^\perp$, because $P_U T f = P_U T P_V f = 0$, and similarly $Tf \in V^\perp$. Consequently, $f \in U$ yields $Tf \in (U^\perp \cap V^\perp) \cap (U+V) = (U+V)^\perp \cap (U+V) = 0$. A similar argument establishes that $Tf = 0$ for $f \in V$. Since $(U+V) + (U+V)^\perp$ is dense in H, T must be the zero operator; hence $P_U P_V = P_V P_U$. Next consider $S = P_U P_V - P_{U \cap V}$. A direct calculation yields:

$$\begin{aligned} S^2 &= (P_U P_V - P_{U\cap V})(P_U P_V - P_{U\cap V}) \\ &= P_U P_V P_U P_V - P_{U\cap V} P_U P_V - P_U P_V P_{UV} + P_{U\cap V}^2 \\ &= P_U P_V - P_{U\cap V} p_U P_V - P_{U\cap V} + P_{U\cap V} = (I - P_{U\cap V}) P_U P_V \end{aligned}$$

Since we obviously have $S = S^*$, the foregoing yields:

$$\begin{aligned} S^2 &= S^* S = (S^* S)^* = (S^2)^* = ((I - P_{U\cap V}) P_U P_V)^* \\ &= P_V P_U (I - P_{U\cap V}) = P_U P_V - P_{U\cap V} = S \end{aligned}$$

Therefore S is an orthogonal projection. If $f \in U \cap V + (U^\perp + V^\perp)$, then $Sf = 0$ (using that $U^\perp + V^\perp \subset (U \cap V)^\perp$). Since the closure of $U^\perp + V^\perp$ is $(U \cap V)^\perp$ and since S is continuous, it follows that $Sf = 0$ for any f in $(U \cap V) + (U \cap V)^\perp$; hence $S = 0$. □

Observe that for any linear subspace of H, U for example, the closure of U in H is given by $U^{\perp\perp}$. The advantage of the analytic proof given above is that we do not have to verify the axioms of a noncommutative topology for the set of finite products of idempotent elements of $\mathcal{L}(H)$. In that way we would arrive at the following result too; we again provide an analytic proof.

Theorem 2.8
Let $\mathcal{P}$ be a family of normal operators acting on the Hilbert space H; that is, $P \in \mathcal{P}$ implies that $PP^* = P^*P$. Suppose that for all finite $\{P_1, \ldots, P_n\} \subset \mathcal{P}$ we have $(P_1 \ldots P_n)^2 = P_1 \ldots P_n$; then for any finite $\{P_1, \ldots, P_n\} \subset \mathcal{P}$ we have $P_1 \ldots P_n = (P_1 \ldots P_n)^* P_{R(P_1 \ldots P_n)}$ where for any operator T, $R(T)$ stands for the range of T; in particular all elements of $\mathcal{P}$ commute with one another.

Proof
First we establish that every $P \in \mathcal{P}$ is idempotent. For $P \in \mathcal{P}$, $(PP^* - P)P^* = P(P^*)^2 - PP^* = PP^* - PP^* = 0$; if $Pf = 0$, then $(PP^* - P)f = P^*Pf - Pf = 0$; thus $PP^* - P$ is zero on $R(P^*) + \ker(P)$. The latter space is dense in H; hence by continuity of P, $P = P^* = PP^* = P^*P = P^2 = P_{R(P)}$. We proceed by induction, supposing the result holds for any finite $\{P_1, \ldots, P_{n-1}\} \subset \mathcal{P}$. Take $P \in \mathcal{P}$, $P \notin \{P_1, \ldots, P_{n-1}\}$, which is thus necessarily an orthogonal projection. The induction hypothesis implies $P_1 \ldots P_{n-1} = (P_1 \ldots P_{n-1})^* = P_{R(P_1 \ldots P_{n-1})}$ and $P_1 \ldots P_n = P_{R(P_1 \ldots P_{n-1})} P_n$, with $P_n = P$, leads to $P_1 \ldots P_n = P_{R(P_1 \ldots P_{n-1})} P_n = (P_{R(P_1 \ldots P_{n-1})} P_n)^* = P_{R(P_1 \ldots P_n)}$. Also if $\{P_1, \ldots, P_{i-1}, P, P_{i+1}, \ldots, P_{n-1}\}$ is considered, then again the same argument implies, with an interchanging of P and P_i, and the claim follows. □

We point out that in general it is not obvious that products of comparable operators may be comparable. We have $P \le Q$ whenever $P = PQ (= QP = PQP)$ for orthogonal projectors. If $P_j \leq Q_j$ for $j \in \{1, \ldots, n\}$, then $P_1 \ldots P_n \leq Q_1 \ldots Q_n$, but equality does not entail $P_j = Q_j$ for $j = 1, \ldots, n$. For suitable operators such a result may be proved; we include an example.

Example 2.3: (J. van Casteren)
If $0 \leq P_j \leq Q_j, j = 1, \ldots, n$ for orthogonal projections P_j, Q_j such that the following hold:

i. $(Q_j \ldots Q_k - P_j \ldots P_k)P_k(Q_k \ldots Q_j - P_k \ldots P_j)$
$\geq (Q_j \ldots Q_k - P_j \ldots P_k)(Q_k \ldots Q_j - P_k \ldots P_j)$, for $1 \leq j \leq k \leq n-1$

ii. $(Q_k \ldots Q_j - P_k \ldots P_j)P_{j-1}(Q_j \ldots Q_k - P_j \ldots P_k)$
$\geq (Q_k \ldots Q_j - P_k \ldots P_j)(Q_j \ldots Q_k - P_j \ldots P_k)$ for $2 \leq j \leq k \leq n$.

Then $Q_1 \ldots Q_n = P_1 \ldots P_n$ if and only if $P_j = Q_j, j = 1, \ldots, n$.

We defined points and quasipoints in Section 2.3. Let us point out some facts in the particular case of $L(H)$.

First, one easily verifies that $L(H)$ has no points; indeed if $[A]$ is a point of $L(H)$ given by its filter $\overline{A}$, then $V\{\mathbb{C}u_\alpha, \alpha \in \mathcal{A}\} = H$ for some selected basis $\{u_\alpha, \alpha \in \mathcal{A}\}$; hence $\mathbb{C}u_\alpha \in \overline{A}$ for some suitable $\alpha \in \mathcal{A}$. If $U \in \overline{A}$, then $U \cap \mathbb{C}u_\alpha \in \overline{A}$; hence $\mathbb{C}u_\alpha \subset U$ because $0 \notin \overline{A}$ by assumption; choose $V \in L(H)$ such that V nor $V^\perp$ contains $\mathbb{C}u_\alpha$; then $V \vee V^\perp = H$ yields that either V or $V^\perp$ is in $\overline{A}$, but that contradicts $\mathbb{C}u_\alpha \notin V, \mathbb{C}u_\alpha \notin V^\perp$. Of course $L(H)$ has minimal points (quasipoints) since maximal filters always exist. We have observed that maximal filters define idempotent elements of $C(L(H))$ (see Lemma 2.4) and if $\lambda \in \overline{A}, \beta \in \overline{A}$, then $\lambda \wedge \beta \in \overline{A}$ follows. Let us also recall that a directed set A in a poset Λ (with 0 and 1) is said to be **pointed** if for all $\lambda \notin A$ there exists a $\mu \in A$ such that $\gamma \leq \lambda, \gamma \leq \mu$ implies $\gamma = 0$.

Proposition 2.32
The minimal points of $L(H)$ are exactly given by the pointed filters. There are two types of pointed filters:

i. $\overline{A} = \{U \in L(H), u_\alpha \in U$ for some $u_\alpha \neq 0$ in $H\}$.

ii. $\overline{A}$ *contains all V of finite codimension in H.*

Proof
If some cofinite dimensional U is not in $\overline{A}$, then there is a $V \in \overline{A}$ such that $U \cap V = 0$; therefore, V is finite dimensional. Thus there must exist a $W \in \overline{A}$ with minimal dimension as such. If $\dim W > 1$, then pick a subspace $W' \subset W$ with $\dim W' = 1$; by assumption $W' \notin \overline{A}$; hence there is a $U' \in \overline{A}$ such that $W' \cap U' = 0$, but that contradicts $U' \supset W$. Consequently $\dim W = 1$ and $\overline{A}$ is as claimed in i. The remaining case ii is exactly the case where all cofinite dimensional V are in $\overline{A}$. Note that in general a pointed filter is maximal; indeed if $\overline{A}$ is pointed and $\overline{A} \subsetneqq \overline{B} \subsetneqq L(H)$, then there is a $V \in \overline{B}$ such that $V \notin \overline{A}$, and thus there exists a $W \in \overline{A}$ such that $W \cap V = 0$, contradicting $W, V \in \overline{B}$ and $\overline{B} \neq L(H)$. Conversely, if $\overline{B}$ is a maximal

filter in $L(H)$ such that $U \notin \overline{B}$, then $A = \{U \cap V, V \in \overline{B}\}$ is a directed set because for $V, W \in \overline{B}$, $U \cap (V \cap W) \subset U \cap V, U \cap W$; hence $\overline{A} \supset \overline{B}$ is a strict inclusion of filters because $U \in \overline{A}$, $U \notin \overline{B}$. The maximality assumption on $\overline{B}$ then implies $0 \in \overline{A}$; therefore $U \cap V = 0$ for some $V \in \overline{B}$ and consequently $\overline{B}$ is pointed. □

The foregoing property of $L(H)$ is shared by more classical types of lattices, for example, topologies.

Proposition 2.33
Let X be a topological space satisfying T_1; write $L(X)$ for the lattice of open subsets of X (sometimes denoted Open(X)). If A is a pointed directed set for $L(X)$, then it is one of three possible types:

i. $\cap\{U \in A\} = \{x\}$ *for some $x \in X$ and every open neighborhood of x in $\overline{A}$.*

ii. $\cap\{U \in A\} = \emptyset$ *and* $\cap\{\overline{U}, U \in A\} = \{x\}$ *for some $x \in X$ (where $\overline{U}$ is the closure of U in X).*

iii. $\cap\{U \in A\} = \emptyset$ *and $X - K \in \overline{A}$ for every closed compact set K in X (compact here means having the finite intersection property).*

Proof
Suppose $x, y \in I = \cap\{U \in L(X)\}$. If $y \neq x$ then, in view of the T_1-property, we may select an open neighborhood V_y of y such that $x \notin V_y$; thus $V_y \notin \overline{A}$. Note that we may replace A by its filter $\overline{A}$ because the pointedness assumption is preserved. Thus, there is a $U \in \overline{A}$ such that $U \cap V_y = \emptyset$; hence $y \notin U$ and then $y \notin I$. It follows that $I = \{x\}$ and the claims in i follow. In the remaining cases we have $I = \emptyset$. Suppose there is a closed compact K such that $X - K \notin \overline{A}$. Since $\overline{A}$ is pointed there is a $V \in A$ such that $V \cap (X - K) = \emptyset$; that is, $V \subset K$ or $\overline{V} \subset K$. Look at $I' = \cap\{\overline{U}, U \in \overline{A}\}$. Since for V as above, $\overline{V}$ is compact, it follows that $I' \neq \emptyset$ unless $\emptyset \in \overline{A}$, a case that may be excluded because $\emptyset \notin A$. If $x \neq y$ are both in I' and V_y is an open neighborhood of y such that $x \notin V_y$, then $U \cap V_y = \emptyset$ for some U in $\overline{A}$, while on the other hand $y \in V \cap \overline{U}$. Because U is dense in $\overline{U}$, this leads to a contradiction unless $y = x$; thus $I' = \{x\}$ and the claims of ii are proved. The remaining case is the one where $X - K \in \overline{A}$ for all closed compact subsets K of X, as stated in iii.

The Stone space, originally constructed for Boolean algebras, has been defined also for arbitrary lattices (I do not recall where this first appeared in the literature), but for us the Stone space, as a set, is nothing but the part of $C(\Lambda)$ corresponding to the pointed directed sets, so this definition extends to noncommutative topologies. It is also clear how to define a **generalized Stone topology** on the above defined set, $SC(\Lambda)$ for example. □

2.7.1 The Generalized Stone Topology

Consider a noncommutative topology Λ and $C(\Lambda)$. For $\lambda \in \Lambda$, let $O_\lambda \subset C(\Lambda)$ be given by $O_\lambda = \{[A], \lambda \in \overline{A}\}$. It is trivial to verify $O_{\lambda\wedge\mu} \subset O_\lambda \cap O_\mu$, $O_{\lambda\vee\mu} \supset O_\lambda \cup O_\mu$, and therefore the O_λ define a basis for a topology on $C(\Lambda)$, called the **generalized Stone topology**. We may restrict attention to the point-spectrum Sp(Λ), or the quasipoint spectrum $QS_p(\Lambda)$; the topology defined on these subsets of $C(\Lambda)$ will again be called the generalized Stone topology. Observe that on the quasispectrum, writing $QO_\lambda = \{[A], [A] \in QS_P(\Lambda), \lambda \in \overline{A}\}$ we actually obtain $QO_{\lambda\wedge\mu} = QO_\lambda \cap QO_\mu$ but still only $QO_{\lambda\vee\mu} \supset QO_\lambda \cup QO_\mu$. On the other hand, writing $PO_\lambda = \{[A], [A] \in S_p(\Lambda), \lambda \in \overline{A}\}$ we have $PO_{\lambda\wedge\mu} \subset PO_\lambda \cap PO_\mu$, $PO_{\lambda\vee\mu} = PO_\lambda \cup PO_\mu$. This follows from the fact that for $[A]$ in Sp(Λ) we do have that $\lambda, \mu \in \overline{A}$ entails $\lambda \wedge \mu \in \overline{A}$; indeed (cf. Definition 2.3.3.). On Sp(Λ) the generalized Stone topology is nothing but the point-topology. On $SP(\Lambda)$, writing $SPO_\lambda = \{[A], [A] \in SP(\Lambda), \lambda \in \overline{A}\}$, we have both equalities: $SPO_{\lambda\wedge\mu} = SPO_\lambda \cap SPO_\mu$ and $SPO_{\lambda\vee\mu} = SPO_\lambda \cup SPO_\mu$.

In the foregoing one may replace Λ by the pattern topology $\mathcal{T}$ (or by $T(\Lambda)$ and similar restrictions Sp$\mathcal{T}$ or SP$\mathcal{T}$ as defined earlier; in all cases we shall use the same label—*generalized topology* or *generalized Stone space*—and it will be clear from the context which one it is. Finally, the generalized Stone topology may also be defined on the commutative shadow $SL(\Lambda)$ (see Proposition 2.1), which is a modular lattice, then of course its induced topology on $QS_P(SL(\Lambda))$ is exactly the Stone topology of the Stone space of $SL(\Lambda)$.

In the special case $\Lambda = L(H)$, the generalized Stone space defined on $QS_P(L(H))$ $= QSP(L(H))$ is exactly the classical Stone space that can be used in Gelfand duality theory for $L(H)$ and $\mathcal{L}(H)$. A word of warning perhaps; since $L(H)$ is not satisfying the weak FDI property, one may not expect a result like Corollary 2.4. In fact, whereas $QSP(L(H)$ is rather big, $S_p(L(H)) = SP(L(H))$ **is empty** (see remarks preceeding Proposition 2.3.2). This fact will have a deeper meaning when we aim to develop some sheaf theory over general Λ. In Section 2.4. the basic properties were introduced and we stressed the transfer from (pre-)sheaves over Λ to (pre-)sheaves over $C(\Lambda)$. This will turn out to be of essence in the case $\Lambda = L(H)$ because there are no sheaves (there is not enough "cohesion" between the element of $L(H)$ if one tries to view them as open sets in some generalized topology) over $L(H)$. There will be many sheaves over $C(L(H))$ allowing sheafification of presheaves; in fact, this will already be possible over $QS_P(L(H))$ (see Chapter 4).

An important notion in Gelfand duality theory for $L(H)$ is the notion of spectral family and of observable function. In the following we will see, to our surprise, that a spectral family is essentially just a separated filtration.

We shall consider a **totally ordered group** Γ in the sequel, however, it would be enough to consider a totally ordered poset with meet and join defined for every subfamily. In applications: $\Gamma \subset \mathbb{R}^n_+$.

Definition 2.9: Γ-Spectral Family

Let Λ be a noncommutative topology; then a Γ**-filtration** of Λ is a family $\{\lambda_\alpha, \alpha \in \Gamma\}$ such that for $\alpha \le \beta$ in Γ, $\lambda_\alpha \le \lambda_\beta$ in Λ and $\vee\{\lambda_\alpha, \alpha \in \Gamma\} = 1$ in Λ (i.e., we consider exhaustive filtrations). A Γ-filtration is **separated** whenever $\gamma = \inf\{\gamma_\alpha, \alpha \in \mathcal{A}\}$ in Γ entails that $\lambda_\gamma = \wedge\{\lambda_{\gamma_\alpha}, \alpha \in \mathcal{A}\}$ in Λ, and $0 = \wedge\{\lambda_\gamma, \gamma \in \Gamma\}$. A Γ**-spectral family** is just a separated Γ-filtration; it may be seen as $F : \Gamma \to \Lambda, \gamma \mapsto \lambda_\gamma$ where F is a poset map with $F(\gamma) = \lambda_\gamma$ satisfying the separatedness condition. Note that, by definition, the order in $\wedge\{\lambda_{\gamma_\alpha}, \alpha \in \mathcal{A}\}$ does not matter while on the other hand the λ_{γ_α} need not be idempotent.

The foregoing definition applied with $\Gamma = \mathbb{R}, +$ and $\Lambda = L(H)$ yields the usual notion of spectral family. A well-known example (connected to the Hamilton operator of the harmonic oscillator) is obtained as follows: let x_n, $n \in \mathbb{N}$ be an orthogonal basis of a separable Hilbert space H and define for $\gamma \in \mathbb{R}$, $L(H)_\gamma = \vee\{\mathbb{C}x_n, n \leq \gamma\}$. This is in fact also an example of a Γ-filtration with discrete support as introduced in [1]. First let us continue with some general facts.

A Γ-spectral family on Λ is said to be **idempotent** if $\lambda_\gamma \in \mathrm{id}_\wedge(\Lambda)$ for every $\gamma \in \Gamma$.

We say that Γ is **indiscrete** if for all $\gamma \in \Gamma, \gamma = \inf\{\tau, \gamma < \tau\}$, for example, $\Gamma = \mathbb{R}^n, +$.

Proposition 2.34
If Γ is indiscrete, then every Γ-spectral family is idempotent.

Proof
Since obviously $\gamma = \inf\{\tau, \gamma \leq \tau\}$ in Γ, we have $\lambda_\gamma = \wedge\{\lambda_\tau, \gamma \leq \tau\}$. Since in the latter expression the order of the λ_τ is irrelevant, we may rephrase this as $\lambda_\gamma = \lambda_\gamma \wedge (\bigwedge_{\gamma<\tau} \lambda_\tau) = \lambda_\gamma \wedge \lambda_\gamma$; consequently $\lambda_\gamma \in \mathrm{id}_\wedge(\Lambda)$. □

Corollary 2.13
If $\Gamma = \mathbb{R}_{,+}$, then every Γ-spectral family is necessarily idempotent.

Proposition 2.35

1. *For any Γ-spectral family on Λ, for any $\gamma, \tau \in \Gamma$, $\lambda_\gamma \wedge \lambda_\tau = \lambda_\tau \wedge \lambda_\gamma = \lambda_\delta$ where $\delta = \min\{\tau, \gamma\}$.*
2. *If the Γ-spectral family is idempotent, then for $\gamma, \tau \in \Gamma$, $\lambda_\gamma \wedge \lambda_\tau = \lambda_\gamma \underset{\cdot}{\wedge} \lambda_\tau$ and the Γ-spectral family on Λ is in fact a Γ-spectral family of $SL(\Lambda)$, the commutative shadow of Λ.*

Proof

1. Since Γ is totally ordered, we may assume that $\gamma < \tau$. From $\lambda_\gamma = \wedge\{\lambda_\delta, \gamma \leq \delta\}$ we obtain $\lambda_\gamma \leq \lambda_\gamma \wedge \lambda_\tau$. Hence, $\lambda_\gamma = \lambda_\gamma \wedge \lambda_\tau$. Similarly, $\lambda_\gamma \leq \lambda_\tau \wedge \lambda_\gamma$ because the order of the λ_δ in $\wedge\{\lambda_\delta, \gamma \leq \delta\}$ is irrelevant; thus $\lambda_\gamma = \lambda_\tau \wedge \lambda_\gamma = \lambda_\gamma \wedge \lambda_\tau$ follows.
2. If the $\lambda_\gamma, \gamma \in \Gamma$, are idempotent and they all commute with one another in view of 1, then $\lambda_\gamma \wedge \lambda_\tau$ is idempotent for all $\gamma, \tau \in \mathcal{T}$, i.e. $\lambda_\gamma \underset{\cdot}{\wedge} \lambda_\tau = \lambda_\gamma \wedge \lambda_\tau$. □

2.7.2 Note

The foregoing results hold for any poset satisfying A.1, ..., A.8; let us fix conventions and call such a structure a noncommutative lattice (knowing that this term would fit certain generalizations of this situation, but we restrict to the case where A.1, ..., A.8. hold here). For any $\mu \in \Lambda$ we have a noncommutative lattice $\Lambda(\mu)$, which as a set is just $\{\lambda \in \Lambda, \lambda \leq \mu\}$ equipped with the induced operations $\wedge, \vee$; if Λ is noncommutative topology, then $\Lambda(\mu)$ is a noncommutative lattice also satifying A.9.

A filtration F on a noncommutative lattice Λ is **right bounded** if $\lambda_\gamma = 1$ for some $\gamma \in \Gamma$; F is **left bounded** if $\lambda_\delta = 0$ for some $\delta \in \Gamma$.

For a right-bounded Γ-filtration $F : \Gamma \to \Lambda$ on a **noncommutative lattice** Λ, we may define for every $\mu \in \Lambda$ the **induced filtration** $F|\mu : \Gamma \to \Lambda(\mu)$, where we use $\mu = 1_{\Lambda(\mu)}$ for the unit element of $\Lambda(\mu)$. The exhaustivity property of $F|\mu$ follows from $\mu_\alpha = \mu \wedge \lambda_\alpha = F|\mu(\alpha)$ for $\alpha \in \Gamma$, and thus $\mu_\gamma = \mu \wedge 1 = \mu$ for $\gamma \in \Gamma$ such that $\lambda_\gamma = 1$. Now when F is a Γ-spectral family of Λ, it is not true that $F|\mu$ is also separated! Indeed, if we look at $\delta = \inf\{\delta_\alpha, \alpha \in \mathcal{A}\}$ in Γ, then $\lambda_\delta = \wedge\{\lambda_{\delta_\alpha}, \alpha \in \mathcal{A}\}$ in Λ, but $\mu \wedge \lambda_\delta$ and $\wedge\{\mu \wedge \lambda_\alpha, \alpha \in \mathcal{A}\}$ need not be equal a priori.

Proposition 2.36
Let F define a right-bounded Γ-spectral family on Λ; then $F|\mu$ is a spectral family of the noncommutative lattice $\Lambda(\mu)$ in each of the following cases:

a. $\mu \in \mathrm{id}_\wedge(\Lambda)$ *and* μ *commutes with all* $\lambda_\alpha, \alpha \in \Gamma$.

b. $\mu \wedge \lambda_\alpha$ *is idempotent for each* $\alpha \in \Gamma$.

Proof

a. Clearly $\wedge\{\mu\wedge\lambda_\alpha, \alpha \in \mathcal{A}\} = \mu\wedge(\wedge\{\lambda_\alpha, \alpha \in \mathcal{A}\}) = \mu\wedge\lambda_\delta$, if $\delta = \inf\{\alpha, \alpha \in \mathcal{A}\}$ in Γ.

b. From $\wedge\{\mu \wedge \lambda_\alpha, \alpha \in \mathcal{A}\} = \mu \wedge \ldots \leq \mu \wedge (\wedge_\alpha\{\lambda_\alpha, \alpha \in \mathcal{A}\}) = \mu \wedge \lambda_\delta$, if $\delta = \inf\{\alpha, \alpha \in \mathcal{A}\}$ in Γ, on one hand, but $\mu \wedge \lambda_\delta \leq \mu \wedge \lambda_\alpha$ for all $\alpha \in \mathcal{A}$ with $\mu \wedge \lambda_\alpha$ idempotent on the other, it follows that $\mu \wedge \lambda_\delta = \wedge\{\mu \wedge \lambda_\alpha, \alpha \in \mathcal{A}\}$, as desired. □

An element μ with property a in the foregoing proposition will be called an F**-centralizer** of Λ.

Corollary 2.14
If Λ is a lattice, then for every $\mu \in \Lambda$, a right-bounded Γ-spectral family of Λ induces a right-bounded spectral family on $\Lambda(\mu)$.

Proof
In a lattice every μ is an F-centralizer. □

If F defines a Γ-spectral family on Λ, a noncommutative lattice, then to $\lambda \in \Lambda$ we may associate $\sigma(\lambda) \in \Gamma \cup \{\infty\}$, $\sigma(\lambda) = \inf\{\gamma, \lambda \leq \lambda_\gamma\}$, where we agree to put $\inf\emptyset = \infty$. The map $\sigma : \Lambda \to \Gamma \cup \{\infty\}$ may be seen as the generalization of the principal symbol degree when filtered rings and their associated graded rings are being considered, so we can expect to obtain the character of a valuation order function.

We refer to σ as the **observable function** of F. Clearly $\sigma(\lambda \wedge \mu) \leq \min\{\sigma(\lambda), \sigma(\mu)\}$, $\sigma(\lambda \vee \mu) \leq \max\{\sigma(\lambda), \sigma(\mu)\}$ and the **domain of** σ is defined as $\cup\{[0, \lambda_\gamma], \gamma \in \Gamma\}$. Note that $\vee\{\lambda_\gamma, \gamma \in \Gamma\} = 1$ does not imply $D(\sigma) = \Lambda$; this may even be checked for $\Gamma = \mathbb{R}_{(+)}$, $\Lambda = L(H)$.

Let $F : P \to L(H)$ be a Γ-spectral family on $L(H)$. For an arbitrary linear subspace V in H we define $\gamma_V \in \Gamma, \gamma_V = \inf\{\gamma \in \Gamma, V \subset L(H)_\gamma\}$, again putting $\inf\emptyset = \infty$. The map $\mathcal{P} : L(H) \to \Gamma \cup \{\infty\}, U \mapsto \mathcal{P}(U) = \gamma_U$ is well defined. It is easily verified that $\mathcal{P}(U + V) \leq \max\{\mathcal{P}(U), \mathcal{P}(V)\}, \mathcal{P}(U \cap V) \leq \min\{\mathcal{P}(U), \mathcal{P}(V)\}$ for U and V in $L(H)$. The function $\mathcal{P}$ allows a function $\underline{\mathcal{P}}$ defined on H by putting $\underline{\mathcal{P}}(x) = \mathcal{P}(\mathbb{C}x)$; we shall simplify notation and just talk about $\mathcal{P}$, called the **pseudo-place** of the Γ-spectral family. Then, any Γ-spectral family defines a function on the **projective Hilbert space** $\mathbb{P}H$, which may be identified to the lines in $H, \overline{\mathcal{P}} : \mathbb{P}H \to \Gamma, \underline{\mathbb{C}v} \mapsto \overline{\mathcal{P}}(\underline{\mathbb{C}v}) = \mathcal{P}(\mathbb{C}v)$, where we wrote $\underline{\mathbb{C}v}$ for the line $\mathbb{C}v$ viewed as an element of $\mathbb{P}H$. Such interpretation may be generalized to situations where the elements of Λ are represented by subobjects of some object in a suitable category, but we see no need to go into this generalization here.

We point out that the pseudo-place aspect of $\mathcal{P}$ translates to the function $\overline{\mathcal{P}}$ in the following sense: for $\mathbb{C}w \subset \mathbb{C}v + \mathbb{C}u$ we have $\overline{\mathcal{P}}(\underline{\mathbb{C}w}) \leq \max\{\overline{\mathcal{P}}(\underline{\mathbb{C}v}), \overline{\mathcal{P}}(\underline{\mathbb{C}u})\}$.

Now let us consider a linear subspace $U \subset H$ such that $\mathbb{P}U \subset \overline{\mathcal{P}}^{-1}(]-\infty, \gamma])$, then for $u \neq 0$ in U we have $\mathcal{P}(\mathbb{C}u) \leq \gamma$, or $u \in L(H)_\gamma$. Hence the largest U in H such that $\mathbb{P}U$ is in $\overline{\mathcal{P}}^{-1}(]-\infty, \gamma])$ is exactly $L(H)_{\underline{\gamma}}$. This states that it is possible to reconstruct the filtration F from the knowledge of $\overline{\mathcal{P}}$.

Classically one looks at $\Gamma = \mathbb{R}, +$, an $\mathbb{R}$-spectral family that will be called a **spectral family** with respect to the quotient topology induced in $\mathbb{P}H$ from H; the map $\overline{\mathcal{P}}$ is lower semicontinuous.

In the classical theory of the quantum lattice $L(H)$ a self-adjoint operator on H determines a unique family of orthogonal projections $P_\gamma, \gamma \in \Gamma = \mathbb{R}$, such that $L(H)_\gamma = P_\gamma H$ defines a spectral family in $L(H)$. Moreover, the original self-adjoint operator can be recovered from his family of spectral projections P_γ by the Riemann-Stieltjens integral $\int_{-\infty}^{+\infty} \gamma d P_\gamma$. Assuming that we started from an operator in $\mathcal{L}(H)$, that is, a bounded operator Q say, then an operator S in $\mathcal{L}(H)$ commutes with Q if and only if S commutes with all the P_γ in the spectral family defining Q. Commuting bounded operators are therefore characterized by the fact that their spectral projections generate a commutative subalgebra in $\mathcal{L}(H)$.

Proposition 2.37
Maximal abelian regular (Von Neumann) subalgebras of $\mathcal{L}(H)$ correspond bijectively with maximal distributive sublattices of $L(H)$.

Proof
We start from the ring theoretical fact that any regular $\mathbb{C}$-subalgebra A of $\mathcal{L}(H)$ is generated by its idempotents, that is, by the projections contained in A. If A is commutative, then these projections commute and therefore $\wedge$ of such coincides with the product of the operators in the algebra $\mathcal{L}(H)$, and the distributivity of the

lattice of projections contained in A follows. On the other hand, it is equally well known and clear that some sublattice of $L(H)$ is distributive exactly when for all U, V in it the projections P_U and P_V commute. A maximal distributive sublattice D of $L(H)$ generates a commutative regular subalgebra of $\mathcal{L}(H)$, $\mathbb{C}(D)$ for example. If $\mathbb{C}(D) \subsetneqq A$, where A is another abelian regular subalgebra of $\mathcal{L}(H)$, then the lattice of projections contained in A strictly contains D and it is distributive in view of the first part of the proof. The latter contradicts the maximality assumption on D; hence $\mathbb{C}(D)$ is maximal as an abelian regular subalgebra of $\mathcal{L}(H)$. □

Since any Γ-spectral family is certainly a directed set they define as elements of $C(\Lambda)$, we call these elements **Γ-points** of Λ, denoting the set of Γ-points by $[\Gamma] \subset C(\Lambda)$. We may think of $[\mathbb{R}] \subset C(L(H))$ as being identified via the Riemann-Stieltjens integral with the set of self-adjoint operators on H.

Let $\sigma : \Lambda \to \Gamma \cup \infty$ be the observable function associated to a Γ-spectral family on Λ defined by $F : \Gamma \to \Lambda$. We define $\widehat{\sigma} : C(\Lambda) \to \Gamma \cup \{\infty\}$, $[A] \mapsto \inf\{\gamma \in \Gamma, \lambda_\gamma \in \overline{A}\}$; the latter is the observable function corresponding to the Γ-filtration on $C(\Lambda)$ defined by $[A]_\gamma$ where $\gamma \in \Gamma$, $[A]_\gamma$ is the class of the smallest filter containing λ_γ, that is, the filter $\{\mu \in \Lambda, \lambda_\gamma \leq \mu\}$. This is clearly a Γ-spectral family because in fact $[A]_\gamma = [\lambda_\gamma]$. We define $[\Gamma] \cap Sp(\Lambda) = \Gamma - Sp(\Lambda)$, $[\Gamma] \cap QSp(\Lambda) = \Gamma - QSp(\Lambda)$, and similarly we may define $\Gamma - SP(\Lambda)$ and $\Gamma - QSP(\Lambda)$. If Γ is indiscrete, for example, if $\Gamma = \mathbb{R}$ then $\Gamma - Sp(\Lambda) = \Gamma - SP(\Lambda)$ and $\Gamma - QSP(\Lambda) = \Gamma - QS_p(\Lambda)$ because then the spectral family is idempotent.

In view of Proposition 2.35(1) a Γ-spectral family is contained in a sublattice (that is, with commutative $\wedge$) of the noncommutative lattice Λ; in fact $\{\lambda_\gamma, \gamma \in \Gamma\}$ is such a lattice. In a noncommutative lattice there exist maximal commutative sublattices (i.e., with respect to the induced structures). Let $Ab(\Lambda)$ be the set of maximal commutative sublattices of Λ. Every Γ-spectral family is a Γ-spectral family in some $B \in Ab(\Lambda)$; that is, every element of $\Gamma - S_p(\Lambda)$, respectively $\Gamma - QS_p(\Lambda)$, appears as an element of $\Gamma - S_p(B)$, respectively $\Gamma - QS_p(B)$. The choice of the notation B should remind us of the term *Boolean sector* to which it will reduce in case $\Gamma = \mathbb{R}, +, \Lambda = L(H)$. This follows easily from the observation that a maximal "commutative" sublattice of the $L(H)$ (in the sense that the corresponding projection operators commute) is in fact a maximal distributive sublattice of $L(H)$. Indeed, if $P_U \wedge P_V = P_V \cap P_U$, then $P_U P_V = P_V P_U$ entails that $P_U P_V$ is idempotent, so Proposition 2.31 yields that $P_U P_V = P_V P_U = P_{U \wedge V}$. Whereas in general it is possible that a Γ-point of Λ appears as a Γ-point of different maximal commutative sublattices of Λ, in case $\Gamma = \mathbb{R}$ and $\Lambda = L(H)$, there is a stringent relation expressed in the following.

Lemma 2.24

Every $[\mathbb{R}]$-point of $QSP(L(H))$ is in $QSP(B)$ for a unique $B \in Ab(L(H))$.

There is a relation between the "size" of $\mathbb{R}$ and the size of $L(H)$ as posets for some undefined notion of size! This aspect is also present in the following.

Observation 2.6

If $\mathcal{P} : H \to \mathbb{R} \cup \{\infty\}$ is the pseudo-place defined on $\mathbb{H}$ by a spectral family, then $\mathcal{P}^{-1}(\mathbb{R})$ is dense in H.

Proof
For $x \in H, \mathcal{P}(x) = \mathcal{P}(\mathbb{C}x) = \inf\{\gamma \in \mathbb{R}, \mathbb{C}x \subset L(H)_\gamma\}$. So if the spectral family is given by $\gamma \mapsto L(H)_\gamma$ for $\gamma \in \mathbb{R}$, then $\mathcal{P}^{-1}(\mathbb{R}) \supset \sum_{\gamma\in\mathbb{R}} L(H)_\gamma$. Since the closure of $\sum_{\gamma\in\mathbb{R}} L(H)_\gamma$ in H is by the exhaustive property of the filtration exactly H, because $1 = \cup_\gamma L(H)_\gamma$, where $1_{L(H)} = H$, the density follows. □

2.7.3 Project: Noncommutative Gelfand Duality

Since the occasion presents itself, one may want to develop a spectral duality theory without passing through the consideration of maximal commutative sublattices of Λ; this would depend on the structure of the algebra $\mathcal{L}(H)$ and not exclusively on its maximal abelian Von Neumann regular subalgebras.

In case of general Γ it is a subproject to continue the investigation of Γ-points of Λ together with their relation to Γ-points of maximal commutative sublattices of Λ.

Note that $L(H)$ is not a noncommutative topology in the strict sense because Axiom A.10 does not hold in $L(H)$ (perhaps the term **skew topology**, already appearing in [46], could be used to refer to structures satisfying A.1., ..., A.9 but not necessarily A.10). As a question of an almost philosophical nature, one may ask how much of a noncommutative Gelfand duality could actually follow from a careful analysis of the semigroup of words in the P_U (including 0 and 1) as elements of $\mathcal{L}(H)$; indeed the relations expressing the equality of different words in the P_U when evaluated in $\mathcal{L}(H)$ essentially contain the basic information on the algebraic and geometric structure of $\mathcal{L}(H)$ and $L(H)$.

Combined with the use of the operator order for the partical order, the foregoing actually rephrases the problem of describing the noncommutative topology, built on the finite words in the projectors, which has $L(H)$ for its commutative shadow. For our use here as an example, the consideration of the lattice $L(H)$ is sufficient. From the point of view of noncommutative topology it is important to further the more general study of "skew topology" properties related to H.

2.8 Ore Sets in Schematic Algebras

Consider any ring R and S a multiplicatively closed subset of R such that $0 \neq S$, $1 \in S$. We say that S is a **left Ore subset** of R if the following conditions hold :

LO.1 For $r \in R, s \in S$ there exist $r' \in R, s' \in S$ such that $s'r = r's$.

LO.2 If $rs = 0$ for some $r \in R, s \in R$, then there exists an $s' \in S$ such that $sr = 0$.

By complete left-right symmetry one may define **right Ore sets**, and we shall refer to an S as before as an Ore set if it is both a left and right Ore set. To a left Ore set S we may correspond a (left) Gabriel filter $\mathcal{L}(S) = \{L$, a left ideal of $R, L \cap S \neq \emptyset\}$. There is a torsion theory κ_S on R-mod defined by saying that $M \in R$-mod is κ_S-torsion, $M = \kappa_S(M)$, if and only if for every $m \in M$ there exists $L \in \mathcal{L}(S)$ such that

$Lm = 0$. The corresponding localization functor is denoted by Q_S and the localization morphism by $j_S : R \to Q_S(R)$. We use terminology and notation as in Section 2.6. For an Ore set S we can form the ring of fractions $S^{-1}R$ obtained by taking equivalence classes of pairs (s, r) with $s \in S$, $r \in R$, with respect to the relation $(s, r) \sim (s', r')$ if and only if there exists $s'' \in S$ such that $s''(s'r - sr') = 0$. In this case the left and right localizations coincide and $Q_S(R) = S^{-1}R$. For $M \in R$-mod we write $\kappa_S(M) = \{m \in M, \exists L \in \mathcal{L}(S)$ such that $Lm = 0\}$ and $\kappa_S(M) = \operatorname{Ker} j_S(M)$, $j_S(M) : M \to Q_S(M)$ being the localizing morphism. It is easy to show, using the flatness of $S^{-1}R$ as a (left) $\mathcal{R}$-module, that Q_S is an exact functor and $Q_S(M) \cong Q_S(R) \otimes M$.

We say that R is **affine schematic** if there exists a finite set of nontrivial Ore sets of R (that is, not consisting of invertible elements of R), say $S_1, \ldots, S_n$, such that for every choice of $s_i \in S_i$, $i = 1, \ldots, n$, we obtain $\sum_{i=1}^{n} Rs_i = R$, or equivalently $\cap_i \mathcal{L}|(S_i) = \{R\}$. In torsion theoretic language, the foregoing just means that $\kappa_{S_1} \wedge \ldots \wedge \kappa_{S_n} = \xi$ where ξ is the trivial kernel functor (which may be seen as corresponding to the torsion theory given by the trivial Ore set $\{1\}$). At the end of this section we list many interesting examples of (affine) schematic algebras.

Now look at a graded algebra $A = K \oplus A_1 \oplus A_2 \oplus \ldots$ that is, a connected positively $\mathbb{Z}$-graded K-algebra, meaning that $A_0 = K$. We put $A_+ = A_1 \oplus A_2 \oplus \ldots$ and write $\mathcal{L}(\kappa_+)$ for the Gabriel filter consisting of left ideals of A containing some A_+^n for some $n \in \mathbb{N}$. The latter corresponds to a kernel functor κ_+ and a torsion theory on A-mod where $\kappa_+(M) = \{m \in M, A_+^n m = 0$ for some $n \in \mathbb{N}\}$. If S is a homogeneous Ore set of A, then we have κ_S on A-mod, but we can also look at a graded kernel functor κ_S^g on the Grothendieck category A-gr with corresponding localization functor Q_S^g : A-g $\to$ A-gr (see Section 3.4 for some facts related to graded localization theory in connection with noncommutative projective spaces). A K-algebra A as before is said to be **schematic** if there exists a finite set of homogeneous Ore sets of A, which are nontrivial in the sense that every S considered has $S \cap A_+ \neq \emptyset$, and satisfying one of the following equivalent properties:

S.1. For $s_i \in S_i$, $i = 1, \ldots, n$ we have $A_+^m \subset \sum_i As_i$ for some $m \in \mathbb{N}$.

S.2. $\cap_i \mathcal{L}(S_i) = \mathcal{L}(\kappa_+)$, (or $\cap_i \mathcal{L}^g(S_i) = \mathcal{L}^g(\kappa_+)$ if one restricts the filter to graded left ideals.

S.3. If $M \in A$-gr $\cap_i \kappa_{S_i}(M) = \kappa_+(M)$.

S.4. We have $\kappa_+ = \kappa_{S_1} \wedge \ldots \kappa \kappa_{S_n}$ (or $\kappa_+^g = \kappa_{S_1}^g \wedge \ldots \wedge \kappa_{S_n}^g$ when graded kernel functors are being considered).

For an arbitrary ring R we let $\mathcal{O}(R)$ be the set of Ore sets of R containing 1 but not 0. In case A is a positively graded ring, then we write $\mathcal{O}^h(A)$ for S as before, which consist of homogeneous elements in the gradation of A, and which are nontrivial in the sense that $S \cap A_+ \neq \emptyset$. In the sequel we continue with the situation of a positively graded A; the ungraded case can be deduced from this by "forgetting" the gradation and obvious modifications. We consider $W(A)$ the set of words in letters from $\mathcal{O}^h(A)$, that is, the free monoid on $\mathcal{O}^h(A)$. If $W = S_1 \ldots S_n \in W(A)$, then we write "$w \in W$" to mean that $w \in A$ has the form $s_1 \ldots s_n$ with $s_i \in S_i$, $i = 1, \ldots, n$ (small letters appear in the same order as the big letters). We form a category $\underline{W}$ by taking the elements of $W(A)$

for the objects while for $W = S_1 \dots S_n$, $W' = T_1 \dots T_m$ we define homomorphisms by putting $\mathrm{Hom}(W', W) = \{W' \to W\}$ or $\emptyset$ depending on whether there exists a strictly increasing mapping α, $\alpha : \{1, \dots, n\} \to \{1, \dots, m\}$, for which $S_i = T_{\alpha(i)}$, or not. By definition $\mathrm{Hom}(W', W)$ is always a singleton if it is not empty. For $M \in A$-gr we define $Q_W(M) = (Q_{S_n} \circ \dots \circ Q_{S_1})(M) = Q_{S_n}(A) \otimes_A \dots \otimes_A Q_{S_1}(A) \dots \otimes_A M$, where the Q_{S_i} are the localization functors corresponding to S_i (we could restrict attention to the graded Q^g_W by using the $Q^g_{S_i}$ and work in A-gr, everything is easily modified). To $W \in \mathcal{W}(A)$ we may associate a filter $\mathcal{L}(W)$ consisting of left ideals of A, given by $\mathcal{L}(W) = \{L, w \in L$ for some $w \in W\}$. One easily checks for $w, w' \in W$ and for some $a, b, \in A$ that we have: $aw = bw' \in W$ and moreover for $w \in W$, $a \in A$ there is $w' \in W$ and $b \in A$ such that $w'a = bw$ (from repeated use of Ore conditions). For $M \in A$-mod, $\kappa_W(M) = \{x \in M, wx = 0$ for some $w \in W\}$ is exactly the kernel of the morphism $M \to Q_W(M)$. The κ_W is an exact preradical on A-mod (and A-gr) but it is not necessarily **idempotent**! The way we defined $\mathcal{L}(W)$ entails that it has a cofinal system of graded left ideals defining $\mathcal{L}^h(W)$ and inducing an exact preradical of A-gr.

The idempotency of κ_W or $\mathcal{L}(W)$ is related to compatibility conditions between the letters in W. The following result, although easy enough to prove, went unnoticed in the Ring Theory literature until we picked it up for its importance in the construction of a noncommutative Grothendieck topology structure on the $\underline{\mathcal{W}}$.

Proposition 2.38
See 2.1.8. in [44]. For (left) Ore sets in any ring R, say S and T, the following properties are equivalent:

C.1. *$\mathcal{L}(TS) = \{L$ left ideal of $R, ts \in L$ for some $t \in T, s \in S\}$ is an idempotent (Gabriel) filter.*

C.2. *$\mathcal{L}(ST) \subset \mathcal{L}(TS)$.*

C.3. *The canonical $Q_S Q_T(R) \to Q_{S\vee T}(R)$ is an R-module isomorphism, where $S \vee T$ is the Ore set generated by $S \cup T$.*

C.4. *For an R-module M, $\kappa_S Q_T(M)$ is a $Q_T(R)$-module.*

C.5. *$Q_S Q_T$ is localization functor on R-mod.*

C.6. *$Q_S Q_T(R)$ is a ring with the ring structure inducing the canoncial R-module structure.*

C.7. *$Q_S Q_T(R)$ is a left $Q_T(R)$-module inducing the canonical R-module structure.*

C.8. *For $M \in R$-mod, $\kappa_{TS}(Q_S Q_T(M)) = 0$.*

C.9. *For $M \in R$-mod, $Q_S Q_T(\kappa_{TS}(M)) = 0$.*

The foregoing properties are also equivalent when S and T are everywhere interchanged. The compatibility of S and T is then equivalent to each of the foregoing statements together with its $S - T$-symmetric version. Recall that for Noetherian R we say that κ_S and κ_T are **compatible** if and only if the functors Q_S and Q_T

commute, if and only if $\kappa_S Q_T = Q_T \kappa_S, \kappa_T Q_S = Q_S \kappa_T$. Note that $W' \to W$ in $\underline{\mathcal{W}}$ yields $\mathcal{L}(W) \subset \mathcal{L}(W')$ and moreover for any $V \in \underline{\mathcal{W}}$ we then also have $W'V \to WV$ as well as $VW' \to VW$ being morphisms in $\underline{\mathcal{W}}$. A finite subset $\{W_i, i \in \mathcal{J}\}$ of objects of $\underline{\mathcal{W}}$ is called a **global cover** if $\cap_{i\in\mathcal{J}}\mathcal{L}(W_i) = \mathcal{L}(\kappa_+)$; the existence of at least one global cover is thus guaranteed by the schematic condition (even by words consisting of just one letter)! For any $W \in \underline{\mathcal{W}}$ we put $\text{Cov}(W) = \{W_i W \to W, i \in \mathcal{J}\}$, that is, the covers induced by global covers. The category $\underline{\mathcal{W}}$ together with the defined covers $\text{Cov}(W)$, $W \in \underline{\mathcal{W}}$ is a noncommutative Grothendieck topology. Indeed G.1 and G.2 are easily verified. For the noncommutative "fibre product" axiom we look at $\{U_i W \to W \to W, i \in \mathcal{J}\} \in \text{Cov}(W)$ and a given $W' \to W$ in $\underline{\mathcal{W}}$, say $W = S_1 \dots S_n$ and $W' = V_1 S_1 V_1 S_2 V_2 \dots S_n V_n$ we find $\{U_v W \times_W W' \to W', i \in \mathcal{J}\} \in \text{Cov}(W')$ where we put $U_i W \times_W W' = U_i W' = U_i V S_1 V_1 S_2 \dots S_n V_n$. Verification of the noncommutative Grothendieck topology axiom is now easy, and it also follows from the following.

Proposition 2.39
If $\{U_i, i \in \mathcal{J}\}$ is a global cover in $\underline{\mathcal{W}}$, then for all $V \in \underline{\mathcal{W}}$ we have $\mathcal{L}(V) = \cap_{i\in\mathcal{J}}\mathcal{L}(U_i V)$.

Proof
That $\mathcal{L} \subset \cap_{i\in\mathcal{J}}\mathcal{L}(U_i V)$ is obvious. For the converse, look at a left ideal L of A, $L \in \mathcal{L}(U_i V)$ for all $i \in \mathcal{J}$, say $u_i v_i \in L$ with $u_i \in U_i, v_i \in V$, for $i \in J$. Since $\mathcal{J}$ is finite, we may find a common multiple for the v_i in V, say $v = a_i v_i \in V$, by repeated use of the Ore conditions for the letters appearing in the word V. Moreover $u_i' a_i \in A u_i$ for some $u_i' \in U_i$. The fact that $\{U_i, i \in J\}$ is a global cover in $\underline{\mathcal{W}}$ yields the existence of an $n \in \mathbb{N}$ such that $A_+^n \in \sum_{i\in J} A u_i$ and therefore $A_+^n v \subset L$. Let T be the first letter appearing in V; then there is $t \in T \cap A_+^n$, but then $tv \in L$ together with $tv \in V$ yields $L \in \mathcal{L}(V)$. □

Of course we can look at the noncommutative topology operations stemming from A-tors on κ_S with $S \in \mathcal{O}^h(A)$ (recall that the topology order is opposite to A-tors!); the topology union would then correspond to the intersection of filters, that is, $\mathcal{L}(\lambda_S \vee \lambda_T) = \mathcal{L}(S) \cap \mathcal{L}(T)$ and the latter is an idempotent filter; therefore, $\vee$ is commutative and we shall arrive at a virtual topology. The topology intersection $\lambda_S \wedge \lambda_T$ corresponds to $\mathcal{L}(ST)$ and that is not necessarily idempotent (in fact Proposition 2.38 describes well enough when $\mathcal{L}(ST)$ is idempotent). The fact that the first nine axioms for a noncommutative topology hold is clear because these hold for A-tors (and A-rig if one restricts to graded localizations). Axiom 10 follows exactly from Proposition 2.39, so $W(A)$ also defines a virtual topology. In this example the affine elements described in Section 3.2 are easily recognized; let us work this through as an example.

Consider a schematic Noetherian K-algebra A as before and suppose that A is generated by A_1 over K, a condition very common in geometric situations where A is a graded epimorphic image of a free algebra with its standard gradation. The localization functor Q_S corresponding to any Ore set of A is an exact functor commuting with direct sums in A-mod; for homogeneous S a similar statement holds for Q_S^g in A-gr. Recall that a $\mathbb{Z}$-graded ring R is said to be **strongly graded** if $R_n R_{-n} = R_0$

for all $n \in \mathbb{Z}$ or equivalently $R_1 R_{-1} = R_{-1} R_1 = R_0$. The interesting property of a strongly graded ring is that the categories R-gr and R_0-mod are isomorphic categories. Since any $\gamma \geq \kappa_S$ is compatible with κ_S, it follows that $\text{Tors}\, Q_S(R) \cong \text{gen}(\kappa_S)$, and $\text{Rig}\, Q_S^g(R) \cong \text{gen}_{\text{rig}}(\kappa_S)$ follows (see also 3.4.2) by restriction to graded localization. If $Q_S^g(R)$ is strongly graded, then the graded localization on $Q_S^g(R)$-gr corresponds to the localizations of $Q_S^g(R)_0$-mod; that is, $\text{Tors}\, Q_S^g(R)_0$ is topologically isomorphic to the noncommutative open corresponding to κ_S, that is, is indeed an "affine" open as it may be seen as the spectrum associated to $Q_S^g(R)_0$. Therefore the desired affineness follows from the following.

Proposition 2.40
If A is as above, $A = K < A_1 >$, then for every homogeneous Ore set $S \in \mathcal{O}(A)$ such that $S \cap A_+ \neq \emptyset$, it follows that $Q_S(A)$ is strongly graded.

Proof
Put $B = Q_S(A)$ with $b \in B_d$ if and only if $sb \in A_{n+d}$ for some $s \in S \cap A_n$. It is harmless to replace A by $A/\kappa_S(A)$; that is, we may assume that $A \hookrightarrow_S (A)$; also note that $Q_S(A) = Q_S^g(A)$. If $Z \in B_0$, then $s_n z \in A_n$ for some $s_n \in S \cap A_n$; that is, $s_n z = \sum_i a_1^{(i)} \dots a_n^{(i)}$ with $a_j^{(i)} \in A_1$, $j = 1 \dots n$. Rewrite this as $z = \sum_i (s_n^{-1} a_1^{(i)} \dots a_{n-1}^{(i)}) a_n^{(i)}$ with $s_n^{-1} a_1^{(i)} \dots a_{n-1}^{(i)} \in B_{-1}$. Thus $z \in B_{-1} A_1 \subset B_{-1} B_1 \in B$, follows, or $B_0 = B_{-1} B_1$. Since $B_{-1} \dots B_{-1}.B_1 \dots B_1 = B_0$ we also obtain $B_{-n} B_n = B_0$ for positive n. In a similar way we derive that $B_n = B_0 A_1^n$ and then $B_n = B_1^n$ follows too. Finally if $z \in B_0$ and $s_m \in S \cap A_m$ is such that $s_m z \in A_m$, then $z = (z s_m) s_m^{-1} \subset B_m B_{-m} = B_1^m B_{-m} \subset B_1 (\mathcal{B}_1^{m-1} B_{-m}) \subset B_1 B_{-1}$, thus $B_0 = B_1 B_{-1}$ as well. □

An extended version of the foregoing result can be found in Proposition 3.16.

Example 2.4
The coordinate ring of quantum 2×2-matrices, $O_q(M_2(\mathbb{C}))$, with $q \in \mathbb{C}$, is schematic and a Noetherian domain. This algebra is generated over $\mathbb{C}$ by elements A, B, C, D subjected to the following relations:

- $BA = q^{-2} AB$
- $CA = q^{-2} AC$
- $BC = CB$
- $DB =^{-2} BD$
- $DC = q^{-2} CD$
- $AD - DA = (q^2 - q^{-2}) BC$

In fact one take S_A, S_B, S_C, S_D respectively generated by the powers of A, B, C, D; the schematic condition can be checked stepwise for the consecutive extensions $\mathbb{C}[B, C][A][D]$, which at each step is given by an Ore extension; that is, $O_q(M_2(\mathbb{C}))$ is an iterated Ore extension.

Example 2.5
Quantum Weyl algebras are schematic. Look at an $n \times n$-matrix $(\alpha_{ij}) = A$ with $\alpha_{ij} \in K^* = K - \{0\}$ and let $\overline{q} = (q_1, \dots, q_n)$ be a row with $q_i \neq 0$ in K. Define $\mathbb{A}_n(\overline{q}, A)$ as

the K-algebra generated by $x_1, \ldots, xn, y_1, \ldots, y_n$ subjected to the following relations: putting $\mu_{ij} = \lambda_{ij} q_i$,

$$\begin{aligned} x_i x_j &= \mu_{ij} x_j x_i \\ x_i y_j &= \alpha_{ji} y_i x_i \\ y_j y_i &= \alpha_{ji} y_i y_j \\ x_j y_i &= \mu_{ij} y_i x_j \\ x_j y_j &= q_j y_j x_j + 1 + \textstyle\sum_{i<j} (q_i - 1) y_i x_i \end{aligned}$$

Again this algebra may be obtained as an iterated Ore extension creating stepwise extensions by consecutively adding to $K, x_1, y_1, x_2, y_2, \ldots, x_n, y_n$.

Example 2.6

The Sklyanin algebra is schematic. Let $S_K(A, B, C)$ be the K-algebra generated by three homogeneous elements X, Y, Z of degree 1, with homogeneous defining relations:

$$\begin{cases} A \times Y + BYX + CZ^2 = 0 \\ AYZ + BZY + CX^2 = 0 \\ AZX + BXZ + CY^2 = 0 \end{cases}$$

A proof for this, using a valuation reduction idea, is given in [44].

Example 2.7: E. Witten's Gauge Algebras for $SU(2)$

Consider the $\mathbb{G}$-algebra W generated by X, Y, Z, subjected to the following relations:

$$\begin{cases} XY + \alpha YX + \beta Y = 0 \\ YZ + \gamma ZY + \delta X^2 + \varepsilon X = 0 \\ ZX + \xi XZ + \eta Z = 0 \end{cases}$$

for any $\alpha, \beta, \gamma, \delta, \varepsilon, \xi, \eta \in \mathbb{C}$. This Witten-algebra (and its associated graded rings as well as the Rees ring with respect to the obvious filtration given in terms of the total degree in X, Y and Z) is schematic.

Example 2.8: Woronowicz's Quantum sl_2

Let $W_q(sl_2)$ be the $\mathbb{G}$-algebra generated by X, Y, Z subjected to the following relations:

$$\begin{cases} \sqrt{q} XZ - \sqrt{q}^{-1} ZX = \sqrt{q + q^{-1}} Z \\ \sqrt{q}^{-1} XY - \sqrt{q} YX = -\sqrt{q + q^{-1}} Y \\ YZ - ZY = (\sqrt{q} - \sqrt{q}^{-1}) X^2 - \sqrt{q - q^{-1}} X \end{cases}$$

where classically $q = \exp\left(\frac{2\pi i}{k+2}\right)$ is the Chern coupling constant.
The algebra, as well as its Rees ring for the the standard filtration, is schematic.

Chapter 3

Grothendieck Categorical Representations

3.1 Spectral Representations

We start from a category $\underline{\mathcal{R}}$ allowing products and coproducts. Typical examples we have in mind are amongst others: the category **Ring** of associative rings with unit, the category $R\text{-gr}_G$ of G-graded associative rings with unit for some group G, the category $\underline{\text{Alg}_k}$ of k-algebras, the category R**-filt** of $\mathbb{Z}$-filtered rings (an interesting non-abelian case), and so forth.

To each object R of $\underline{\mathcal{R}}$ we associate a Grothendieck category $\text{Rep}(R)$. For every $f \in \text{Hom}_{\underline{\mathcal{R}}}(S, R)$, $f : S \to R$, we are given an exact functor $f^o = F : \text{Rep}(R) \to \text{Rep}(S)$, which commutes with products and coproducts and satisfies the following conditions:

i. $(1_R)^o = I_{\text{Rep}(R)}$ for every $R \in \underline{\mathcal{R}}$

ii. For $g : T \to S$, $f : S \to R$ in $\underline{\mathcal{R}}$, $(f \circ g)^o = g^o \circ f^o$. We did not demand that for $R \neq S$ in $\underline{\mathcal{R}}$ necessarily $\text{Rep}(R) \neq \text{Rep}(S)$

If $\underline{\mathcal{G}}$ is the class consisting of objects $\text{Rep}(R)$, $R \in \underline{\mathcal{R}}$, we let $\text{Hom}_{\underline{\mathcal{G}}}(\text{Rep}(R), \text{Rep}(S))$ consist of functors of type h^o provided these go from $\text{Rep}(R)$ to $\text{Rep}(S)$. Note that if Rep is separating objects of $\underline{\mathcal{R}}$, then we may write $\text{Hom}_{\underline{\mathcal{G}}}(\text{Rep}(R), \text{Rep}(S)) = \text{Hom}_{\underline{\mathcal{R}}}(S, R)^o$. In any case $\underline{\mathcal{G}}$ as defined above becomes a category.

Definition 3.1
A **Grothendieck categorical representation** (a GC representation) is a contravariant functor $\text{Rep} : \underline{\mathcal{R}} \to \underline{\mathcal{G}}$ commuting with arbitrary products, associating to $f : S \to R$ an exact functor $f^o = F : \text{Rep}(R) \to \text{Rep}(S)$ commuting with products and coproducts.

Several obvious examples come to mind, for example, representing noncommutative rings by their category of left modules, groups by their G-modules, and graded algebras by their categories of graded modules. Such examples exhibit stronger properties than those used in the general definition. This is mainly due to the fact that the objects of $\underline{\mathcal{R}}$ appear in some form also in the representing Grothendieck category, such as a ring as a left module over itself, and so forth. This may be formalized in the following definition.

Definition 3.2
A GC representation $\underline{\mathcal{R}} \to \underline{\mathcal{G}}$ is said to **measure** $\underline{\mathcal{R}}$ if for every $R \in \underline{\mathcal{R}}$ we have an object ${}_R R$ in Rep(R) such that to any $f : S \to R$ in $\underline{\mathcal{R}}$ there corresponds a morphism $f^* :_S S \to \text{Rep}(f)({}_R R) =_S R$, satisfying:

μ.1. If $f = 1_S$ then f^* is the identity of ${}_S S$ in Rep(S).

μ.2. Consider $g : T \to S$, $f : S \to R$ in $\underline{\mathcal{R}}$; then we have :

$$(f \circ g)^* = \text{Rep}(g)(f^*) \circ g^* :_T T \to_T S \to_T R = \text{Rep}(g)\text{Rep}(f)({}_R R)$$

A GC representation Rep : $\underline{\mathcal{R}} \to \underline{\mathcal{G}}$ is said to be **full** if to an epimorphism $\pi : S \to R$ in $\underline{\mathcal{R}}$ there corresponds a full functor Rep(π) : Rep(R) $\to$ Rep(S). A GC representation is **faithful** if for $f : S \to R$ in $\underline{\mathcal{R}}$, Rep(f) is faithful. An $R \in \underline{\mathcal{R}}$ is **Rep-Noetherian** when Rep(R) is a Grothendieck category having a Noetherian generator. Similarly, $R \in \mathcal{R}$ is **locally Rep-Noetherian** whenever Rep(R) has a family of Noetherian generators.

The relation between a GC representation and suitable topologies will be obtained from the hereditary torsion theories existing on the Grothendieck categories. In Section 2.6 we introduced general torsion theory in Grothendieck categories, but here we modify the notation somewhat in order to fit the notation fixed in the introduction of Grothendieck representations. For an arbitrary Grothendieck category $\underline{\mathcal{M}}$ we let Tors($\underline{\mathcal{M}}$) be the set of hereditary torsion theories on $\underline{\mathcal{M}}$; we know that Tors($\underline{\mathcal{M}}$) is a modular lattice with respect to inf and sup of torsion theories (cf. the note following Proposition 2.11), but the operation "product" in the lattice of preradicals $\mathcal{Q}(\underline{\mathcal{M}})$ is noncommutative.

Torsion theories of $\underline{\mathcal{M}}$ will be denoted by $\sigma, \tau, \kappa, \ldots$; then $\mathcal{T}_\sigma$, $\mathcal{F}_\sigma$ will denote the torsion, respectively the torsion-free class of σ. We write $T_\sigma : \underline{\mathcal{M}} \to \underline{\mathcal{M}}$ for the corresponding torsion functor (kernel functor) and $(\underline{\mathcal{M}}, \sigma)$ for the quotient category together with the canonical functors $i_\sigma : (\underline{\mathcal{M}}, \sigma) \to \underline{\mathcal{M}}$, $a_\sigma : \underline{\mathcal{M}} \to (\underline{\mathcal{M}}, \sigma)$ (see Theorem 2.17). Then $i_\sigma a_\sigma = Q_\sigma$ is the localization functor $\underline{\mathcal{M}} \to \underline{\mathcal{M}}$ associated to σ.

For $R \in \underline{\mathcal{R}}$ we abbreviate TorsRep(R) to Top(R).

In case $\underline{\mathcal{R}}$ has an initial object, k say, then we call Top(k) the **initial space** for Rep($\underline{\mathcal{R}}$).

To a morphism $f : S \to R$ in $\underline{\mathcal{R}}$ we have associated a functor $F = \text{Rep}(R) \to \text{Rep}(S)$. Since F is exact and commutes with coproducts, it defines a map F^o : Top(S) $\to$ Top(R); indeed, if $\gamma \in$ Top(S) we may define $F^o(\gamma)$ by taking for $\mathcal{T}_{F^o(\gamma)}$ the class of objects X in Rep(R) such that $F(X) \in \mathcal{T}_\gamma$; when F derives from f we shall write $\widetilde{f}$ for F^o.

Definition 3.3
A faithful Grothendieck representation that measures $\underline{\mathcal{R}}$ is said to be **spectral** if for all $A \in \underline{\mathcal{R}}$, $\gamma \in$ Top(A) and $\tau \in$ gen(γ) we are given the following:

i. An object $A(\gamma)$ in $\underline{\mathcal{R}}$ together with a morphism $f_\gamma : A \to A(\gamma)$ such that the morphism f_γ^* in Rep(A), $f_\gamma^* :_A A \to \text{Rep}(f_\gamma)({}_{A(\gamma)}A(\gamma))$ is exactly the localization morphism ${}_A A \to Q_\gamma(A)$.

ii. A morphism $f_{\tau_\gamma} : A(\gamma) \to A(\tau)$ in $\underline{\mathcal{R}}$ fitting into a commutative diagram:

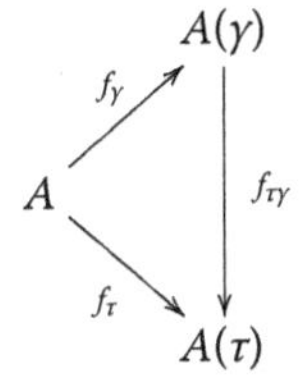

iii. If $\xi_o(A)$ stands for the trivial element of Top(A), that is, the zero element of the latticc Tors(Rep(A)), then $A = A(\xi_o(A))$.

Proposition 3.1
Consider an exact functor F, F : Rep(S) $\to$ Rep(R), commuting with direct sums; then F^o : Top(R) $\to$ Top(S) has the following properties:

i. *For $\gamma \le \sigma$ in* Top(R)*, $F^o(\gamma) \le F^o(\sigma)$ in* Top(S).

ii. *If $U \subset$* Top(R)*, then $F^o(\wedge U) = \wedge F^o(U)$.*

iii. *For τ in* Top(S)*, let ξ_τ be the minimal torsion theory having all $F(T_\tau)$, $T_\tau \in \mathcal{T}_\tau$; in its torsion class, in other words $\mathcal{T}_{\xi_\tau}$ is the torsion class generated by the $F(T_\tau)$, $T_\tau \in \mathcal{T}_\tau$. We have $(F^o)^{-1}(\text{gen}(\tau)) = \text{gen}(\xi_\tau)$.*

Proof
An easy exercise.

In general, for $U \subset \text{Top}(A) : \text{gen}(\wedge U) \supset U\{\text{gen}(\tau), \tau \in U\}$, $\text{gen}(\vee U) = \cap\{\text{gen}, \tau \in U\}$; moreover there is a trivial torsion theory ξ defined by $T_\xi = 0$, and a maximal torsion theory χ defined by $T_\chi = \text{Rep}(A)$. Consequently, the sets gen(τ), $\tau \in$ Top(A) define a topology on Top(A). □

Corollary 3.1
The gen-topology. *F^o as in Proposition 3.1 is continuous in the gen-topology.*

For γ in Top(A) *we have the reflector a_γ* : Rep(A) $\to$ (Rep(A), γ) *and the associated map a^o_γ* : Tors(Rep(A), γ) $\to$ Top(A). *Note that the forgetful functor* □ : (Rep(A), γ) $\to$ Rep(A) *is not exact in general, so it does not yield an associated map* Top(A) $\to$ Tors(Rep(A), γ).

In the part of this section we only ***consider GC representations that are faithful****. The categories considered are assumed to have a zero object.*

Proposition 3.2
Suppose that Rep is spectral and $\gamma \in$ Top(A) *for A in $\underline{\mathcal{R}}$; take $\tau \in$* gen(γ) *and write $\tilde{\tau}$ for $\widetilde{f_\gamma}(\tau) \in$* Top($A(\gamma)$) *and $f_{\tilde{\tau}}$ for the corresponding morphism in $\underline{\mathcal{R}}$, $f_{\tilde{\tau}} : A(\gamma) \to A(\gamma)(\tilde{\tau})$, which exists because of the spectral property of Rep applied to $A(\gamma)$ in $\underline{\mathcal{R}}$.*

Then we have:

i. $\text{Rep}(f_{\tau_\gamma})({}_{A(\tau)}A(\tau)) = Q_{\tilde{\tau}}({}_{A(\gamma)}A(\gamma))$.

ii. $\text{Rep}(f_\gamma)(Q_{\tilde{\tau}}({}_{A(\gamma)}A(\gamma))) = Q_\tau({}_A A)$.

Proof

i. By the spectral property of Rep, there exists an $A(\gamma)(\tilde{\tau})$ in $\underline{\mathcal{R}}$ together with a morphism $f_{\tilde{\tau}} : A(\gamma) \longrightarrow A(\gamma)(\tilde{\tau})$ with corresponding morphism in $\text{Rep}(A(\gamma))$, $f_{\tilde{\tau}}^*$ say, $f_{\tilde{\tau}}^* :_{A(\gamma)} A(\gamma) \longrightarrow \text{Rep}(f_{\tilde{\tau}})({}_{A(\gamma)(\tilde{\tau})}A(\gamma)(\tilde{\tau}))$, which is exactly the localization homomorphism in $\text{Rep}(A(\gamma))$, ${}_{A(\gamma)}A(\gamma) \longrightarrow Q_{\tilde{\tau}}({}_{A(\gamma)}A(\gamma))$.

On the other hand we may consider:

$$\text{Rep}(f_\gamma)(f_{\tilde{\tau}}^*) : Q_\gamma({}_A A) \to \text{Rep}(f_\gamma)(Q_{\tilde{\tau}}({}_{A(\gamma)}A(\gamma))),$$

the latter being equal to

$$\text{Rep}(f_\gamma)\text{Rep}(f_{\tilde{\tau}})({}_{A(\gamma)(\tilde{\tau})}A(\gamma)(\tilde{\tau})) = \text{Rep}(f_{\tilde{\tau}} f_\gamma)({}_{A(\gamma)(\tilde{\tau})}A(\gamma)(\tilde{\tau})).$$

From the composition $A \underset{f_\gamma}{\longrightarrow} A(\gamma) \underset{f_{\tilde{\tau}}}{\longrightarrow} A(\gamma)(\tilde{\tau})$, we obtain:

$$\begin{array}{lcc} \text{Rep}(f_\gamma)(f_{\tilde{\tau}}^*)f_\gamma^* :_A A \longrightarrow & & \text{Rep}(f_{\tilde{\tau}} f_\gamma)({}_{A(\gamma)(\tilde{\tau})}A(\gamma)(\tilde{\tau})) \\ * & & \downarrow \cong \\ & & \text{Rep}(f_\gamma)(Q_{\tilde{\tau}}({}_{A(\gamma)}A(\gamma))) \end{array}$$

Now ${}_{A(\tau)}A(\tau)$ in $\text{Rep}(A(\tau))$ is such that the localization morphism ${}_A A \to Q_\tau(A)$ is exactly given by $f_\tau^* :_A A \longrightarrow \text{Rep}(_\tau)$ $({}_{A(\tau)}A(\tau))$. Consider $M_\tau(\gamma)$ in $\text{rep}(A(\gamma))$ defined as follows:

$$M_\tau(\gamma) = \text{Rep}(f_{\tau_\gamma})({}_{A(\tau)}A(\tau))$$

and let $f_{\tau_\gamma}^* :_{A(\gamma)} A(\gamma) \longrightarrow M_\tau(\gamma)$ be the morphism in $\text{Rep}(A(\gamma))$ obtained from the measuring property of Rep. Obviously, $\tilde{\tau}(M_\tau(\gamma)) \subset M_\tau(\gamma)$ in $\text{Rep}(A(\gamma))$. The definition of $\tilde{\tau}$ yields: $\text{Rep}(f_\gamma)(\tilde{\tau}M_\tau(\gamma))$ is τ-torsion in $\text{Rep}(A)$, and moreover, the exactness of $\text{Rep}(f_\gamma)$ yields $\text{Rep}(f_\gamma)(\tilde{\tau}M_\tau(\gamma) \subset \text{Rep}(f_\gamma)(M_\tau(\gamma)) = \text{Rep}(f_\tau)({}_{A(\tau)}A(\tau))$, where the latter is τ-torsion free because it equals $Q_\tau({}_A A)$. Thus $\text{Rep}(f_\gamma)(\tilde{\tau}M_\tau(\gamma))$ $= 0$ and the faithfulness of Rep then implies that $\tilde{\tau}M_\tau(\gamma) = 0$. It follows that we may factorize $f_{\tau\gamma}^* |_{A(\gamma)}A(\gamma) \to M_\tau(\gamma)$, via $B_\tau(\gamma) =_{A(\gamma)} A(\gamma)/\tilde{\tau}({}_{A(\gamma)}A(\gamma))$. So we have the following sequence in $\text{Rep}(A)$:

$${}_A A \underset{f_\gamma^*}{\longrightarrow} Q_\gamma({}_A A) \underset{\text{Rep}(f_\gamma)(f_{\tau_\gamma}^*)}{\longrightarrow} Q_\tau({}_A A)$$

It is therefore clear that $\mathrm{Rep}(f_\gamma)(M_\tau(\gamma)/B_\tau(\gamma))$ is τ-torsion in $\mathrm{Rep}(A)$ because the co-kernel of $\mathrm{Rep}(f_\gamma)(f^*_{\tau_\gamma})$ is τ-torsion, then $M_\tau(\gamma)/B_\tau(\gamma)$ is $\tilde{\tau}$-torsion in $\mathrm{Rep}(A(\gamma))$. It follows that $M_\tau(\gamma) \subset Q_{\tilde{\tau}}({}_{A(\gamma)}A(\gamma))$ in $\mathrm{Rep}(A(\gamma))$. The definition of $Q_{\tilde{\tau}}({}_{A(\gamma)}A(\gamma))$ makes it $\tilde{\tau}$-torsion over $B_\tau(\gamma)$ in $\mathrm{Rep}(A(\gamma))$; consequently $\mathrm{Rep}(f_\gamma)(Q_{\tilde{\tau}}({}_{A(\gamma)}(A(\gamma)))$ is contained in $Q_\tau({}_AA)$ as it is τ-torsion over ${}_AA/\tau({}_AA)$. Since $M_\tau(\gamma) \subset Q_{\tilde{\tau}}({}_{A(\gamma)}A(\gamma))$, the functor $\mathrm{Rep}(f_\gamma)$ takes the value $Q_\tau(A)$ for both objects, so the faithfulness and exactness of $\mathrm{Rep}(f_\gamma)$ entails that $M_\tau(\gamma) = Q_{\tilde{\tau}}({}_{A(\gamma)}A(\gamma))$. This establishes i above. Observe also that (*) entails that the morphisms $\mathrm{Rep}(f_\gamma)(f^*_{\tilde{\tau}})f^*_\gamma$ and $\mathrm{Rep}(f_\gamma)(f^*_{\tau_\gamma})f^*_\gamma$ are the same.

ii. This follows from i by applying $\mathrm{Rep}(f_\gamma)$ to both members and then again applying the exactness and faithfulness of $\mathrm{Rep}(f_\gamma)$. □

Corollary 3.2

i. *Consider* $\delta \leq \tau, \gamma \leq \tau$ *in* $\mathrm{Top}(A)$ *and* $\tilde{\tau}_1 \in \mathrm{Top}(A(\delta)), \tilde{\tau}_2 \in \mathrm{Top}(A(\gamma))$ *constructed as before (we prefer to write* τ_1, τ_2 *rather than* τ_δ, τ_γ*). We obtain the following commutative diagram of morphisms in* $\underline{\mathcal{R}}$*:*

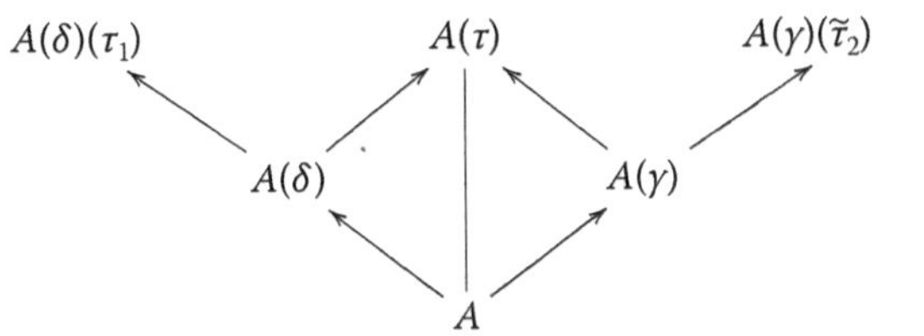

ii. *Consider the following objects:*

$M_1 =_{A(\delta)(\tilde{\tau}_1)} A(\delta)(\tilde{\tau}_1)$ in Rep $(A(\delta)(\tilde{\tau}_1))$
$M_2 =_{A(\gamma)(\tilde{\tau}_2)} A(\gamma)(\tilde{\tau}_2)$ in Rep $(A(\gamma)(\tilde{\tau}_2))$
$M =_{A(\tau)} A(\tau)$in Rep $(A(\tau))$

Then the following relations hold:

a. $\mathrm{Rep}(f_{\tilde{\tau}_1})(M_1) = Q_{\tilde{\tau}}({}_{A(\delta)}A(\delta)) = \mathrm{Rep}(f_{\tau_\delta})(M)$

b. $\mathrm{Rep}(f_{\tilde{\tau}_2})(M_2) = Q_{\tilde{\tau}}({}_{A(\gamma)}A(\gamma)) = \mathrm{Rep}(f_{\tau_\gamma})(M)$

c. $\mathrm{Rep}(f_\delta)\mathrm{Rep}(f_{\tilde{\tau}_1})(M_1) = Q_\tau({}_AA) = \mathrm{Rep}(f_\gamma)\mathrm{Rep}(f_{\tilde{\tau}_2})(M_2)$
$= \mathrm{Rep}(f_\tau)({}_{A(\tau)}A(\tau) = \mathrm{Rep}(f_\tau)(M)$

Proof
Apply the foregoing result (twice). □

The foregoing may be compared to the classical fact that for a ring R and $\tau \geq \gamma$ in Tors(R−mod) we have $Q_\tau(R) = Q_{\tilde{\tau}} Q_\gamma(R)$; in our abstract setting Rep($A(\gamma)$) replaces $Q_\gamma(R)$-mod. This shows that we have traced exactly the property of a GC representation, that is, spectrality, necessary to extend the foregoing classical fact to the general categorical situation.

Note that $A(\delta)(\widetilde{\tau}_1)$ and $A(\gamma)(\widetilde{\tau}_2)$ in the foregoing corollary need not be isomorphic in $\underline{\mathcal{R}}$ (the relations in the corollary sum up what we do know).

Write Top$(A)^o$ for the opposite lattice of Top(A). We would like to consider the functor ${}_AP$: $\overline{\text{Top}(A)} \to$ Rep(A), $\tau \to Q_\tau({}_AA)$ as a **structural presheaf** (or in fact a sheaf) for ${}_AA$ with values in Rep(A). Exactly the spectral property of Rep would then allow us to "realize" this structure sheaf in $\underline{\mathcal{R}}$ by considering P : $\overline{\text{Top}(A)} \to \underline{\mathcal{R}}$, $\tau \to A(\tau)$ with structure morphism $f_\tau : A \to A(\tau)$. It is clear that $P(\xi_0(A)) = A$ with $I_A : A \to A$ as the structure morphism. Moreover, for $\gamma \leq \tau$ in Top(A) we take $f_{\tau_\gamma} : A(\gamma) \to A(\tau)$ for the restriction morphism from γ to τ (in Top$(A)^o$ the partial order is reversed when viewing γ and τ in Top(A) as "opens"). For P to be a presheaf we do need an extra property for Rep!

Definition 3.4

A spectral Grothendieck representation Rep is said to be **schematic** if for every triple $\gamma \leq \tau \leq \delta$ in Top(A), for every A in $\underline{\mathcal{R}}$, we have a commutative diagram in $\underline{\mathcal{R}}$:

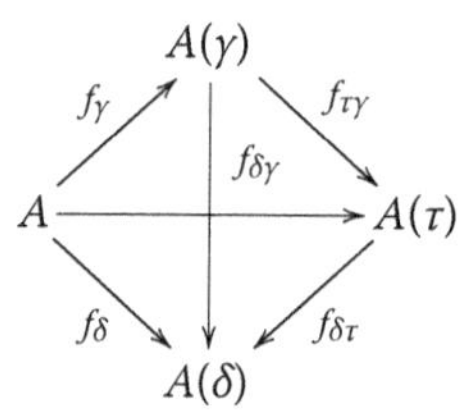

Proposition 3.3

If Rep is schematic, then with notation as before, P : Top(A) $\to \underline{\mathcal{R}}$ *is a presheaf with values in* $\underline{\mathcal{R}}$ *over the lattice* Top$(A)^o$, *for every* A *in* $\mathcal{R}$.

Proof

The composition property of sections follows from $f_{\delta_\gamma} = f_{\delta_\tau} f_{\tau_\gamma}$, and the claim follows easily. □

Note that for a schematic GC representation Rep, the structure presheaf obtained in Proposition 3.3 is constructed via localizations in the representing Grothendieck categories described as in Proposition 3.2.

In the foregoing we have restricted attention to Tors(Rep(A)); that is, we considered a lattice in the usual sense; hence this should be viewed as the commutative shadow of a suitable noncommutative theory. For A in $\underline{\mathcal{R}}$ we shall write $\mathcal{Q}(A)$ for the set of preradicals (see remark before Definition 2.6; note Observation 2.4 too). **Warning**, in earlier work we (and several other authors) have approached hereditary torsion theory via radicals; that is, via the opposite $\mathcal{Q}(A)^{\text{op}}$, the notion $\mathcal{Q}(A)$ expresses the **topology aspects** of the theory more directly!

Applying definitions (e.g., 2.6) and properties of preradicals derived in Section 2.6 to the Grothendieck category $\underline{\mathcal{C}} = \text{Rep}(A)$, we obtain the complete lattice $\mathcal{Q}(A)$ and a duality expressed by an order-reversing bijection: $(-)^{-1} : \mathcal{Q}(A) \to \mathcal{Q}((\text{Rep}(A))^o)$. First let us point out that $(\text{Rep}(A))^o$ is not a Grothendieck category! It is additive and has a projective generator; moreover, it is known to be a **varietal category** (also called *triplable*) in the sense that it has a projective regular generator P, it is co-complete and has kernel pairs with respect to the functor $\text{Hom}(P, -)$, and moreover every equivalence relation in the category is a kernel pair. A concrete description of the opposite of a Grothendieck category is given by U. Oberst ("Duality Theory for Grothendieck Categories and Linearly Compact Rings," *J. Algebra* 15, 1970, 473–542) but sacrificing the varietal aspect for a topological approach. General localization techniques can be developed via the Eilenberg-Moore category of a triple (S. MacLane, *Categories for the Working Mathematician*, Springer-Verlag, Berlin, New York, 1971 [28]). The latter depends on the so-called comparison functor constructed via $\text{Hom}(P, -)$ as a functor to the category of sets. It works well for the category of set-valued sheaves over a Grothendieck topology. As yet we have not investigated whether the approach via the Eilenberg-Moore category remains valid in the case of a noncommutative Grothendieck topology; this may be an interesting project of abstract value.

Now $(-)^{-1}$ defined as an order-reversing bijection between idempotent radicals on $\text{Rep}(A)$ and $(\text{Rep}(A))^o$, we write $(\text{Top}(A))^{-1}$ for the image of $\text{Top}(A)$ in $\mathcal{Q}((\text{Rep}(A))^o)$. This is encoded in the exact sequence in $\text{Rep}(A)$:

$$0 \longrightarrow \rho(M) \longrightarrow M \longrightarrow \rho^{-1}(M) \longrightarrow 0$$

(reversed in $(\text{Rep}(A))^o$). By restricting attention to hereditary torsion theories (kernel functors) when defining $\text{Tors}(-)$, we introduce an asymmetry that breaks the duality because $\text{Top}(A)^{-1}$ is not in $\text{Tors}((\text{Rep}(A))^{\text{op}})$. Write $TT(\underline{\mathcal{G}})$ for the complete lattice of torsion theories (not necessarily hereditary) of the category $\underline{\mathcal{G}}$; then $(TT(\underline{\mathcal{G}}))^{-1} \cong TT(\underline{\mathcal{G}}^{\text{op}})$. Hence we may view $\text{Tors}(\underline{\mathcal{G}})^{-1}$ as a complete sublattice of $TT(\underline{\mathcal{G}}^{\text{op}})$.

For preradicals ρ_1 and ρ_2 in $\mathcal{Q}(\underline{\mathcal{G}})$ we have defined the lattice operations $\rho_1 \wedge \rho_2$, $\rho_1 \vee \rho_2$, as well as the product $\rho_1\rho_2$. Inspired by the duality (see remarks preceding Definition 2.6 and those following Proposition 2.11) we define $\rho_1 \prod \rho_2 = \rho_1 : \rho_2$. Hence an object M of $\underline{\mathcal{G}}$ is $\rho_1 \prod \rho_2$-torsion if and only if there is a subobject $N \subset M$ such that N is ρ_1-torsion and M/N is ρ_2-torsion. The notation $\prod$ suggests that it is topologically an intersection, but as a preradical $\rho_1 \prod \rho_2$ is larger than ρ_1 and ρ_2. Let us denote the similar operation but defined in $\mathcal{Q}((\mathcal{G})^{\text{op}})$ by $\prod^o$; for $\sigma, \tau \in \mathcal{Q}(\underline{\mathcal{G}})$ we let $\sigma \coprod \tau$ be the preradical such that $(\sigma \coprod \tau)^{-1} = \tau^{-1} \prod^o \sigma^{-1}$. The notation $\coprod$ suggests that it is topologically a (noncommutative) union.

Proposition 3.4
With notation as above:

a. *For $\sigma, \tau \in \mathcal{Q}(\underline{G})$, $\mathcal{T}_{\sigma\tau} = \mathcal{T}_\sigma \cap \mathcal{T}_\tau$. Clearly $\sigma\tau \leq \sigma \wedge \tau$. If $\sigma \wedge \tau$ is idempotent, then $\sigma\tau = \tau\sigma = \sigma \wedge \tau = \tau \wedge \sigma$. In particular, when σ and τ are left exact preradicals, then $\sigma\tau = \tau\sigma = \sigma \wedge \tau$, and this is a left exact preradical. Also if*

τ is left exact and σ is idempotent, then $\sigma\tau$ is idempotent and $\sigma\tau = (\sigma \wedge \tau)^o$ (notation of Theorem 2.3(v)).

b. *If σ and τ are idempotent preradicals, then $\sigma \prod \tau$ and $\tau \prod \sigma$ are idempotent preradicals. When only σ is idempotent, then $(\sigma \prod \tau)^o = \sigma \prod \tau^o$. If σ and τ are left exact preradicals, then $\sigma \prod \tau$ and $\tau \prod \sigma$ are left exact preradicals too. Note also that $\mathcal{F}_{\sigma \prod \tau} = \mathcal{F}_\sigma \cap \mathcal{F}_\tau$ (but $\sigma \prod \tau$ is not determined by $\mathcal{F}_{\sigma \prod \tau}$).*

Proof

a. The statements are obvious; let us establish the last one. Let τ be left exact, σ idempotent. For any M in $\underline{\mathcal{G}}$ we have $\sigma\tau(M) = \sigma(\tau(M))$, hence a subobject of $\tau(M)$; therefore the left exactness of τ implies that $\sigma\tau(M)$ is in the τ-pretorsion class. Then $\sigma\tau(\sigma\tau(M)) = \sigma(\sigma\tau(M)) = \sigma\tau(M)$; hence $\sigma\tau$ is an idempotent preradical. Therefore $(\sigma \wedge \tau)^o \leq \sigma\tau$. As observed before, $\sigma\tau(M)$ is in $\mathcal{T}_\sigma$ as well as in $\mathcal{T}_\tau$ (τ is left exact); thus $\sigma\tau \leq \sigma \wedge \tau$ is clear; the definition of $(\sigma \wedge \tau)^o$ then implies that $\sigma\tau = (\sigma \wedge \tau)^o$.

b. Let σ and τ be idempotent preradicals. By definition $(\sigma \prod \tau)(M) = N$, the largest subobject of M such that $M \supset N \supset \sigma(M)$ with $N/\sigma(M)$ being τ-pretorsion. Since σ is idempotent $\sigma(N) = \sigma(M)$ hence $(\sigma \prod \tau)(N) = N_1$, the largest subobject of N, $N \supset N_1 \supset \sigma(N) = \sigma(M)$ such that $N_1/\sigma(M)$ is τ-pretorsion. However, the latter implies $N_1 = N$ in view of the foregoing. In case σ is idempotent but not necessarily τ, then by the foregoing $\sigma \prod \tau^o$ is idempotent, hence $\sigma \prod \tau^o \leq (\sigma \prod \tau)^o$. Since both idempotent preradicals correspond to the same pretorsion class, it follows that $\sigma \prod \tau^o = (\sigma \prod \tau)^o$.

In case both σ and τ are left exact, hence certainly idempotent, we have that $\sigma \prod \tau$ (and $\tau \prod \sigma$) is idempotent. Hence it suffices to check that $\mathcal{T}_{\sigma \prod \tau}$ is a hereditary class (the result for $\tau \prod \sigma$ follows by symmetry). If $M \in \mathcal{T}_{\sigma \prod \tau}$, then $M/\sigma(M)$ is τ-pretorsion; consider a subobject M' of M in $\mathcal{G}$. Since σ is left exact, $M' \cap \sigma(M) = \sigma(M')$. Thus $M'/\sigma(M')$ is isomorphic to a subobject of $M/\sigma(M)$ and therefore the left exactness of τ entails that $M'/\sigma(M') \in \mathcal{T}_\tau$ or $M' \in \mathcal{T}_{\sigma \prod \tau}$. Consequently $\sigma \prod \tau$ is a left exact preradical. □

Proposition 3.5
With notation as before:

a. *If $\sigma, \tau \in \mathcal{Q}(\underline{\mathcal{G}})$ are radicals, then $\sigma\tau$ is a radical; when τ is a radical, then for any $\sigma \in \mathcal{Q}(\underline{\mathcal{G}})$, $(\sigma\tau)^c = \sigma^c\tau$.*

b. *If σ, τ are radicals, then $\sigma \coprod \tau$ and $\tau \coprod \sigma$ are radicals.*

If moreover σ and τ are kernel functors, then $\sigma \coprod \tau = \tau \coprod \sigma = \sigma \wedge \tau (= \sigma\tau = \tau\sigma$ in view of Proposition 3.4).

Proof

a. The statements in a follow by duality from the statements in Proposition 3.4a.

b. The first statement follows by duality from the first statement in Proposition 3.4b. Before dealing with the second statement, let us describe $\mathcal{F}_{\sigma \coprod \tau}$;

now that we are considering radicals σ, τ and $\sigma \coprod \tau$ is a radical too, they are determined by their pre-torsion-free classes. By dualization of the definition of $\prod^o$ in $\mathcal{Q}(\mathcal{G}^{op})$, we observe that an object of $\underline{\mathcal{G}}$, M say, is in $\mathcal{F}_{\sigma \coprod \tau}$ if it fits in a $\underline{\mathcal{G}}$-exact sequence:

$$0 \longrightarrow N \longrightarrow M \longrightarrow M/N \longrightarrow 0$$

with $N \in \mathcal{F}_\sigma$ and $M/N \in \mathcal{F}_\tau$. First let us check that $\sigma \coprod \tau$ is a torsion theory, that is, that $\mathcal{F}_{\sigma \coprod \tau}$ is closed under extensions. So, suppose we are given the following exact sequences in $\underline{\mathcal{G}}$:

$$0 \longrightarrow M_1 \longrightarrow M \underset{\pi}{\longrightarrow} M_2 \longrightarrow 0$$
$$0 \longrightarrow N_1 \longrightarrow M_1 \underset{\pi_1}{\longrightarrow} M_1/N_1 \longrightarrow 0$$
$$0 \longrightarrow N_2 \longrightarrow M_2 \underset{\pi_2}{\longrightarrow} M_2/N_2 \longrightarrow 0$$

where N_1 and N_2 are in $\mathcal{F}_\sigma$, M_1/N_1 and M_2/N_2 are in $\mathcal{F}_\tau$. View N_1 as a subobject of M and let $X \subset M$ be the subobject such that $X/N_1 = \tau(M/N_1)$. Observe that $X \cap M_1 = N_1$ because $X \cap M_1/N_1$ is isomorphic to a subobject of X/N_1, and hence a τ-torsion object because τ is a hereditary torsion theory radical, while on the other hand it is isomorphic to a subobject of M_1/N_1, hence τ-torsion free. Since $\pi(X)$ is a quotient of X/N_1 it must be τ-torsion. But then $\pi_2(\pi(X)) = 0$; hence $\pi(X) \subset N_2$, and we obtain an exact sequence in $\underline{\mathcal{G}}$:

$$0 \longrightarrow N_1 = X \cap M_1 \longrightarrow X \underset{\text{res}\pi}{\longrightarrow} \pi(X) \longrightarrow 0$$

with N_1 and $\pi(X)$ in $\mathcal{F}_\sigma$. Consequently, $X \in \mathcal{F}_\sigma$ because it is closed under extensions. We also obtain the exact sequence:

$$0 \longrightarrow M_1/N_1 \longrightarrow M/X \underset{\gamma}{\longrightarrow} M_2/N_2 \longrightarrow 0$$

where γ is a factorization of $\pi_2\pi$ obtained from $\pi(X) \subset N_2$.

Since M_1/N_1 and M_2/N_2 are in $\mathcal{F}_\tau$, so must be M/X. In the exact sequence in $\underline{\mathcal{G}} : 0 \longrightarrow X \longrightarrow M \longrightarrow M/X \longrightarrow 0$ we now have $X \in \mathcal{F}_\sigma$ and $M/X \in \mathcal{F}_\tau$, thus $M \in \mathcal{F}_{\sigma \coprod \tau}$. This establishes that $\sigma \coprod \tau$ is a torsion theory. We can go on and establish that $\sigma \coprod \tau$ is left exact and that $\sigma \coprod \tau$ is left exact and that $\sigma \coprod \tau = \tau \coprod \sigma$, but all of this will follow if we establish directly that $\sigma \coprod \tau = \sigma \wedge \tau$ ($\wedge$ denoting the lattice operation in the lattice of idempotent preradicals, being the same as the one in the lattice of idempotent radicals, that is, torsion theories). Since $\sigma \wedge \tau$ and $\sigma \coprod \tau$ are radicals, it suffices to establish that $\mathcal{F}_{\sigma\wedge\tau} = \mathcal{F}_{\sigma \coprod \tau}$. Start with $M \in \mathcal{F}_{\sigma \coprod \tau}$ and assume $(\sigma \wedge \tau)(M) \neq 0$; put $X = (\sigma \wedge \tau)(M) \subset \tau(M)$. We know there exists an exact sequence in $\underline{\mathcal{G}} : 0 \longrightarrow N \longrightarrow M \underset{\gamma}{\longrightarrow} M/N \longrightarrow 0$, with $N \in \mathcal{F}_\sigma$ and $M/N \in \mathcal{F}_\tau$. Since τ is left exact, $X \subset \tau(M)$ entails that $X \in \mathcal{T}_\tau$, hence $\gamma(X) \in \mathcal{T}_\tau$, but as $\gamma(X) \subset M/N$ this entails $\gamma(X) = 0$ or $X \subset N$. Then $X \subset \sigma(M)$ leads to X being σ-torsion; hence $X = 0$ as $N \in \mathcal{F}_\sigma$, that is, $(\sigma \wedge \tau)(M) = 0$.

Conversely, start from $M \in \mathcal{F}_{\sigma\wedge\tau}$. If $\sigma(M) \neq 0$, then $\tau(\sigma(M)) = 0$ because otherwise $\tau\sigma(M) \neq 0$ would be $\sigma \wedge \tau$-torsion (Proposition 3.4a.), $\sigma\tau$ and $\tau\sigma$ are $\leq \sigma \wedge \tau$) in M, a contradiction. Of course $M/\sigma(M)$ is in $\mathcal{F}_\sigma$ because σ is a radical and so, from the exact sequence in $\underline{\mathcal{G}}$:

$$0 \longrightarrow \sigma(M) \longrightarrow M \longrightarrow M/\sigma(M) \longrightarrow 0$$

with $\sigma(M) \in \mathcal{F}_\tau$ and $M/\sigma(M) \in \mathcal{F}_\sigma$, $M \in \mathcal{F}_{\sigma \coprod \tau}$ follows. Combining both parts yields $\mathcal{F}_{\sigma \coprod \tau} = \mathcal{F}_{\sigma\wedge\tau}$.

In the last part of the proof, symmetry has obviously been broken; indeed, for nonhereditary torsion theories we cannot use the proof above to arrive at commutativity of $\coprod$ on torsion theories. The problem is that for (nonhereditary) torsion theories we do not know whether $\tau^{-1} \prod^{\circ} \sigma^{-1}$ is radical. Since $\mathcal{F}_{\tau^{-1}\prod^{o}\sigma^{-1}} = \mathcal{F}_{\tau^{-1}} \cap \mathcal{F}_{\sigma^{-1}}$ duality implies that $\mathcal{T}_{\sigma \coprod \tau} = \mathcal{T}_\sigma \cap \mathcal{T}_\tau$; hence if $\sigma \coprod \tau$ is idempotent, or if $\sigma\tau$ or $\tau\sigma$ is idempotent (note that $\mathcal{T}_{\sigma\tau} = \mathcal{T}_\sigma \cap \mathcal{T}_\tau$ too!), then $\sigma \coprod \tau = \sigma\tau = \tau\sigma = \sigma \wedge \tau$. In particular, when we are only interested in left exact preradicals we might define $\sigma \coprod \tau = \sigma\tau$ from the beginning, neglecting the duality with the (noncommutative) topological intersection in $\mathcal{Q}(\underline{\mathcal{G}}^{\mathrm{op}})$.

In the philosophy of the pattern topology as in Chapter 2 we are interested in bracketed expressions involving $\coprod$, $\prod$, and the $\prod$-idempotent elements (hence radicals). So we look at $\mathcal{Q}_h(\underline{\mathcal{G}})$, the set of hereditary preradicals of $\underline{\mathcal{G}}$; this is a subset of $\mathcal{Q}(\underline{\mathcal{G}})$ closed under $\coprod$ and $\prod$ (and $\coprod$ agrees with the preradical product and is a commutative operation in $\Lambda^o = \mathcal{Q}_h(\underline{\mathcal{G}})$). Clearly $i_{\prod}(\mathcal{Q}_h(\underline{\mathcal{G}}))$ consists of the left exact radicals, that is, the kernel functors. Conversely, bracketed expressions $p(\prod, \coprod, \sigma_i)$ (notation of Section 2.2. see after Lemma 2.12) with $\sigma_i \in \mathrm{id}_{\prod}(\Lambda)$, always yield left exact preradicals, hence it is in complete agreement with our interest in noncommutative topologies generated by their intersection-idempotent elements, to look at $\mathcal{Q}_h(\underline{\mathcal{G}})$. We write T for the set of left exact preradicals obtained as finite bracketed expressions as defined above. Are Λ and T now noncommutative topologies, in fact topologies of virtual opens? This follows from the following easy lemma.

Lemma 3.1

With notation and conventions as before:

1. *Λ and T, have* 1 *and* 0 *(the zero preradical, respectively the identity functor)*
2. *If σ, τ, γ are left exact preradicals, then $(\sigma\tau)(\gamma) = \sigma(\tau\gamma)$.*
3. *If $\sigma \leq \tau$ and γ are left exact preradicals, then $\sigma\gamma \leq \tau\gamma, \gamma\sigma \leq \gamma\tau$.*
4. *If σ is a left exact preradical such that $\sigma^n = 0$, then $\sigma = 0$ (observe that $\mathcal{T}_{\sigma^n} = \mathcal{T}_\sigma$).*
5. *If $\sigma \prod \sigma \prod \ldots \prod \sigma = 1$, then $\sigma = 1$.*
6. *If $\sigma \leq \tau$ and γ are left exact preradicals, then we have: $\gamma \prod \sigma \leq \gamma \prod \tau, \pi \prod \gamma \leq \tau \prod \gamma$.*
7. *For left exact preradicals σ, τ, γ, $(\sigma \prod \tau) \prod \gamma = \sigma \prod (\tau \prod \gamma)$.*

8. *For* σ, τ, γ *as before:* $(\sigma \prod \tau)\gamma = (\sigma\gamma \prod \tau)\gamma$.
9. *For* $\sigma, \tau\gamma$ *as before:* $(\sigma \prod \tau)\gamma = \sigma\gamma \prod \tau\gamma$ *whenever* γ *is a radical, that is, in* $\mathrm{id}_{\prod}(\Lambda)$.
10. $\sigma\tau \prod \sigma = \sigma = \sigma \prod \tau\sigma$, *whenever* σ *is radical.*
11. $\sigma(\sigma \prod \tau) = \sigma = (\tau \prod \sigma)\sigma$.
12. *For left exact preradicals* $\sigma\tau = \tau\sigma$ *(see earlier).*
13. Λ *satisfies the FDI property as defined after Definition 1.9, (hence* Λ *satisfies axiom A.10 as defined before Definition 1.9, as well as axiom VOT.3 as defined at the beginning of Section 1.2).*

Proof
All properties follow easily from the preradical calculus; only 13 may need a little explanation. Recall that $\sigma \leq \tau$ is a focused relation if $\sigma \prod \tau = \tau \prod \sigma = \sigma$ (note that we are using $\leq$ in Λ, which is **poset opposite** to $\mathcal{Q}_h(\underline{\mathcal{G}})$); the FDI property holds if for a focused relation $\sigma \leq \lambda$ with $\lambda = \lambda_1 \coprod \lambda_2$ we have $\sigma = (\sigma \prod \lambda_1) \coprod (\sigma \prod \lambda_2)$. Now since $\sigma \leq \lambda$ is focused, we have $\mathcal{T}_\sigma = \mathcal{T}_{\sigma \prod \lambda} = \mathcal{T}_{\lambda \prod \sigma}$ and because $\lambda = \lambda_1 \coprod \lambda_2$ we obtain $\mathcal{T}_\lambda = \mathcal{T}_{\lambda_1} \cap \mathcal{T}_{\lambda_2}$. Therefore, we obtain $\mathcal{T}_\sigma = \mathcal{T}_{\sigma \prod \lambda} \cap \mathcal{T}_{\sigma \prod \lambda} \subset \mathcal{T}_{\sigma \prod \lambda_1} \cap \mathcal{T}_{\sigma \prod \lambda_2}$. For the converse, look at an object M in the latter. That is, we have:

$$M \supset M_1 \supset 0 \text{ with } M_1 \text{ being } \lambda_1\text{-torsion and } M/M_1 \text{ being } \sigma\text{-torsion}$$
$$M \supset M_2 \supset 0 \text{ with } M_2 \text{ being } \lambda_2\text{-torsion and } M/M_2 \text{ being } \sigma\text{-torsion}$$

Look at $M \supset M_1 \cap M_2 \supset 0$; since we are considering left exact preradicals, $M_1 \cap M_2$ is in $\mathcal{T}_{\lambda_1} \cap \mathcal{T}_{\lambda_2} = \mathcal{T}_\lambda$, while on the other hand $M/M_1 \cap M_2$ is σ-torsion (it embeds in $M/M_1 \oplus M/M_2$). We arrive at $M \subset \mathcal{T}_{\sigma \prod \lambda} = \mathcal{T}_\sigma$. □

Corollary 3.3
Λ is a noncommutative topology; T is a topology of virtual opens. The commutative shadow of Λ (and then also of T) is Tors($\underline{\mathcal{G}}$) *with its usual lattice operations* ($\wedge = \prod$, $\vee = \coprod$, *in the notation of Proposition 2.1).*

Returning to the setting of Grothendieck representations, we have for every object A of $\underline{\mathcal{R}}$, the Grothendieck category Rep(A) *and the set of preradicals* $\mathcal{Q}(A)$ *of* Rep(A) *containing* Top(A) = Tors(Rep(A)). *The foregoing construction of $\Lambda(A)$ in $\mathcal{Q}_h(A)$ leads to a noncommutative topology $\Lambda(A)$ having* Top(A) = $\mathrm{id}_{\prod}(\Lambda(A))$ *with its canonical lattice structure for the commutative shadow. The behavior of $\Lambda(A)$ with respect to morphisms $B \to A$ in $\underline{\mathcal{R}}$ would be satisfactory if we can generalize Proposition 3.1 because then a GC representation leads to canonical topologization of the objects of $\underline{\mathcal{R}}$ by noncommutative topologies having the classical lattice of kernel functors for its commutative shadows. A noncommutative space (spectrum) could then be understood as a localization functor on the level of* Rep(A)*-valued sheaves on $\Lambda(A)$, using structure sheaves, with natural transforms between these functors corresponding to morphisms in the base category $\underline{\mathcal{R}}$.*

To a morphism $f : S \to R$ in $\underline{\mathcal{R}}$ we have associated a functor $F = \mathrm{Rep}(f) : \mathrm{Rep}(R) \to \mathrm{Rep}(S)$, which is exact and commutates with coproducts. Define a map $\widehat{F} : \mathcal{Q}_{\mathrm{id}}(S) \to \mathcal{Q}_{\mathrm{id}}(R)$, where $\mathcal{Q}_{\mathrm{id}}$ denotes the lattice of idempotent preradicals by associating to $\gamma \in \mathcal{Q}_{\mathrm{id}}(S)$ the idempotent preradical $\widehat{F}(\gamma)$ defined by the pretorsion class of objects X in $\mathrm{Rep}(R)$ such that $F(X) \in \mathcal{T}_\gamma$. When F derives from f, we shall also write $\widehat{f}$ for $\widehat{F}$ in order to highlight the connection.

Lemma 3.2
With notation as introduced above: $\widehat{F}$ maps $\mathcal{Q}_h(S)$ to $\mathcal{Q}_h(R)$.

Proof
Take $\gamma \in \mathcal{Q}_h(S)$ and look at $M \in \mathcal{T}_{\widehat{F}(\gamma)}$ and a subobject N of M. Thus $F(M) \in \mathcal{T}_\gamma$ and because F is exact we have $F(N) \subset F(M)$; the fact that γ is hereditary then yields that $F(N) \in \mathcal{T}_\gamma$ and hence $N \in \mathcal{T}_{\widehat{F}(\gamma)}$ by definition of $\widehat{F}(\gamma)$. □

Proposition 3.6
Consider any functor $F : \mathrm{Rep}(R) \to \mathrm{Rep}(S)$ such that F is exact and commutes with coproducts; then $\widehat{F}$ has the following properties:

1. *If we restrict $\widehat{F} : \mathcal{Q}_h(S) \to \mathcal{Q}_h(R)$ to $\mathrm{Top}(S)$, then we obtain F^o, as defined before Definition 3.3.*
2. *$\widehat{F}$ is a poset morphism.*
3. *If $\mathcal{F} \subset \mathcal{Q}_h(S)$, then $\widehat{F}(\wedge_{\varphi}\{\varphi \in \mathcal{F}\}) = \wedge_{\varphi}\{\widehat{F}(\varphi), \varphi \in \mathcal{F}\}$.*

Proof
Easy enough. □

On $\mathcal{Q}_h(A)$ for any A in $\underline{\mathcal{R}}$ we define the **gen-topology** in formally the same way as it was introduced on $\mathrm{Top}(A)$, that is, for $\rho \in \mathcal{Q}_h(A)$ we let $\mathrm{gen}(\rho) = \{\tau \in \mathcal{Q}_h(A), \rho \leq \tau\}$. For any $\mathcal{F} \subset \mathcal{Q}_h(A)$ we have:

$$\mathrm{gen}(\wedge\{\varphi, \varphi \in \mathcal{F}\}) \supset \cup\{\mathrm{gen}(\varphi), \varphi \in \mathcal{F}\},$$
$$\mathrm{gen}(\vee\{\varphi, \varphi \in \mathcal{F}\}) = \cap\{\mathrm{gen}(\varphi), \varphi \in \mathcal{F}\}$$

Together with the existence of a minimal left exact preradical ξ and a maximal one χ, the foregoing relations do establish that the sets $\mathrm{gen}(\varphi)$, $\varphi \in \mathcal{Q}_h(A)$ generate (the open sets of) a topology on $\mathcal{Q}_h(A)$. This topology induces on $\mathrm{Top}(A)$ the gen-topology of $\mathrm{Top}(A)$ (see also Corollary 3.1).

Proposition 3.7
In the situation of Proposition 3.6

1. *For $\rho \in \mathcal{Q}_h(R)$, $(\widehat{F})^{-1}(\mathrm{gen}(\rho)) = \mathrm{gen}(\xi_\rho)$, where ξ_ρ is the preradical corresponding to the hereditary pretorsion class generated by the $F(T_\rho)$, $T_\rho \in \mathcal{T}_\rho$.*
2. *$F^\wedge : \mathcal{Q}_h(S) \to \mathcal{Q}_h(R)$ is continuous in the gen-topology.*

Proof
Straightforward (like Proposition 3.1). □

With respect to the operations $\prod$ and $\coprod$, we have the following rules for the behavior of $\widehat{F}$.

Proposition 3.8
Consider an exact functor, commuting with coproducts: $F : \text{Rep}(R) \to \text{Rep}(S)$*, and let* $\widehat{F} : \mathcal{Q}_h(S) \to \mathcal{Q}_h(R)$ *be the corresponding poset map. For* σ, τ *in* $\mathcal{Q}_h(S)$ *we have:*

$$\widehat{F}(\sigma)\coprod \widehat{F}(\tau) = \widehat{F}\left(\sigma \coprod \tau\right),$$
$$\widehat{F}(\sigma)\prod \widehat{F}(\tau) \leq \widehat{F}\left(\sigma \prod \tau\right)$$

Proof
The property with respect to $\coprod$ follows from Proposition 3.6(3) (in fact it extends to $\coprod$ of a family $\mathcal{F} \subset \mathcal{Q}_h(S)$) and the fact that $\coprod$ coincides with $\wedge$ on left exact preradicals. If M in Rep(R) is a pretorsion object for $\widehat{F}(\sigma)\prod \widehat{F}(\tau)$, then we have an exact sequence in Rep(R):

$$0 \longrightarrow \widehat{F}(\sigma)(M) \longrightarrow M \longrightarrow M/\widehat{F}(\sigma)(M) \longrightarrow 0$$

where $M/\widehat{F}(\sigma)(M)$ is $\widehat{F}(\tau)$-pretorsion. By exactness of F we then obtain an exact sequence in Rep(S):

$$0 \longrightarrow F(\widehat{F}(\sigma)(M)) \longrightarrow F(M) \longrightarrow F(M)/F(\widehat{F}(\sigma)(M)) \longrightarrow 0$$

where $F(\widehat{F}(\sigma)(M))$ is σ-pretorsion by definition of $\widehat{F}(\sigma)$, and $F(M)/F(\widehat{F}(\sigma)(M))$ is τ-pretorsion by definition of $\widehat{F}(\tau)$ (and exactness of F).
Consequently, $F(M) \in \mathcal{T}_{\sigma \prod \tau}$ or M is $\widehat{F}(\sigma \prod \tau)$-pretorsion.

In $\mathcal{Q}_h(S)^{\text{op}}$ we have $\widehat{F}(\sigma \prod \tau) \leq \widehat{F}(\sigma)\prod \widehat{F}(\tau)$; hence for an arbitrary morphism $f : S \to R$ we need not obtain a map $\widehat{F}$ taking $\Lambda(S)$ to $\Lambda(R)$. This is related to the fundamental problem concerning functoriality, also appearing in noncommutative geometry (scheme theory for associative algebras). Certain morphisms yield better behavior of $\widehat{F}$, e.g. when f in an epimorphism in $\underline{\mathcal{R}}$ with Rep$(f) = F$ being a full functor.

Definition 3.5
Given a morphism $f : S \to R$ in $\underline{\mathcal{R}}$, then $\sigma \in \mathcal{Q}_h(S)$ is said **to center** f if $\sigma F = F\widehat{F}(\sigma)$ (composition of functors written in antiorder of application): Rep$(R) \to$ Rep(S), $M \to \sigma F(M) = F(\widehat{F}(\sigma)(M))$. The set of all σ in $\mathcal{Q}_h(S)$ that center f will be denoted by $Z_h(f)$. For $Z_n(f) \cap \text{Top}(S)$ we write $Z_{\text{top}}(f)$ and $Z_h(f) \cap \Lambda(S) = Z_\Lambda(f)$.

Corollary 3.4
With notation as above:

1. *If* $\sigma \in Z_h(f)$*, then for any* τ *in* $\mathcal{Q}_h(S)$ *we have:* $\widehat{F}(\sigma)\prod \widehat{F}(\tau) = \widehat{F}(\sigma \prod \tau)$, $\widehat{F}(\sigma) \vee \widehat{F}(\tau) = \widehat{F}(\sigma \vee \tau)$.

2. *If* $\sigma, \tau \in Z_h(f)$, *then we obtain the following equalities:* $\widehat{F}(\sigma)\prod\widehat{F}(\tau) = \widehat{F}(\sigma\prod\tau)$, $\widehat{F}(\tau)\prod\widehat{F}(\sigma) = \widehat{F}(\tau\prod\sigma)$, $\widehat{F}(\sigma)\vee\widehat{F}(\tau) = \widehat{F}(\sigma\vee\tau)$.

Proof

Clearly 2 follows from 1 so let us prove 1. We know from Proposition 3.8 that $\widehat{F}(\sigma)\prod\widehat{F}(\tau) \leq \widehat{F}(\sigma\prod\tau)$. Consider M in Rep(R) that is $\widehat{F}(\sigma\prod\tau)$-pretorsion, that is, $F(M) \in \mathcal{T}_{\sigma\prod\tau}$. We arrive at the existence of an exact sequence in Rep(S):

$$0 \longrightarrow \sigma(F(M)) \longrightarrow F(M) \longrightarrow F(M)/\sigma(F(M)) \longrightarrow 0$$

where $F(M)/\sigma(F(M))$ is τ-pretorsion.

Since $\sigma \in Z_h(f), \sigma(F(M)) = F(\widehat{F}(\sigma)(M))$. Therefore we have that $F(M)/\sigma(F(M)) = F(M/\widehat{F}(\sigma)(M))$ and consequently $M/\widehat{F}(\sigma)(M)$ is $\widehat{F}(\tau)$-pretorsion. From the exact sequence in Rep(R):

$$0 \longrightarrow \widehat{F}(\sigma(M)) \longrightarrow M \longrightarrow M/\widehat{F}(\sigma)(M) \longrightarrow 0$$

we may conclude that M is in the pretorsion class for $\widehat{F}(\sigma)\prod\widehat{F}(\tau)$. The statement concerning the commutative operation $\vee$ follows in a similar way. □

Proposition 3.9

For any morphism, $f : S \to R$, $Z_h(f)$ *is a lattice with respect to* $\wedge(=\coprod)$ *and* $\vee$. *Moreover,* $\prod$ *is inner in* $Z_h(f)$.

Proof

Using Corollary 3.4 we obtain: for $\sigma, \tau \in Z_h(f)$, $(\sigma \wedge \tau)(F) = \sigma\tau F = \sigma(\tau F) = \sigma(F\widehat{F}(\tau)) = \sigma F\widehat{F}(\tau) = \widehat{F}(\sigma)\widehat{F}(\tau) = \widehat{F}(\sigma) \wedge \widehat{F}(\tau)$. Now consider M in Rep(R); then we have an exact sequence in Rep(R):

$$0 \longrightarrow \widehat{F}(\sigma)(M) \longrightarrow \left(\widehat{F}(\sigma)\prod\widehat{F}(\tau)\right)(M) \longrightarrow \widehat{F}(\tau)(M/\widehat{F}(\sigma)(M)) \longrightarrow 0.$$

By applying the exact functor F we obtain an exact sequence in Rep(S), where we put $X = F(\widehat{F}(\sigma)\coprod\widehat{F}(\tau))(M)$ in $F(M)$,

$$0 \longrightarrow \sigma F(M) \longrightarrow X \longrightarrow \tau F(M/\widehat{F}(\sigma)(M)) \longrightarrow 0, \text{ or}$$
$$0 \longrightarrow \sigma F(M) \longrightarrow X \longrightarrow \tau(F(M)/\sigma F(M)) \longrightarrow 0$$

By definition of $\prod$, $(\sigma\prod\tau)(F(M))$ is defined by the exact sequence:

$$0 \longrightarrow \sigma F(M) \longrightarrow \left(\sigma\prod\tau)(F(M)\right) \longrightarrow \tau(F(M)/\sigma F(M)) \longrightarrow 0$$

It follows that $X = (\sigma\prod\tau)(F(M)$. This leads to the equalities: $F(\widehat{F}(\sigma)\prod\widehat{F}(\tau)) = F\widehat{F}(\sigma\prod\tau) = (\sigma\prod\tau)F$. The statements with respect to $\vee$ follow in a similar way. □

Corollary 3.5
$Z_h(f)$, as well as $Z_{\text{top}}(f)$, $Z_\Lambda(f)$, have the structure of a noncommutative topology induced by the corresponding structure on $\mathcal{Q}_h(S)$.

We have before avoided working with filters of left ideals in a ring in order to try to describe the noncommutative intersection in case of a ring A and torsion theories on A-mod. Since the preradicals of interest appear as compositions of torsion theories, it may be interesting anyway to mention a few technical facts.

First, let us stay in the generality of a given Grothendieck category $\underline{\mathcal{G}}$. For torsion theories τ and κ on $\underline{\mathcal{G}}$, we define a class $\mathcal{T}_{\kappa\tau}$ as the class of objects M in $\underline{\mathcal{G}}$ such that there is a subobject N of M; N is κ-torsion and M/N is τ-torsion, or equivalently $M/\kappa M$ is τ-torsion. The class $\mathcal{T}_{\kappa\tau}$ is closed for taking subobjects, direct sums, and images but not necessarily closed under extensions. Of course $\mathcal{T}_{\tau\kappa} \supset \mathcal{T}_\kappa, \mathcal{T}_\tau$ and similar for $\mathcal{T}_{\kappa\tau}$. Let us rephrase some of our statements about the closure operator (avoiding the terminology of Proposition 3.4 and following) and let $\mathcal{T}^{\text{id}}_{\kappa\tau}$ be the closure of $\mathcal{T}_{\kappa\tau}$ under extensions; then $\mathcal{T}^{\text{id}}_{\kappa\tau}$ is a torsion class for a hereditary torsion theory on $\underline{\mathcal{G}}$. The following lemma is along the lines of earlier observations.

Lemma 3.3
With notations as above, $\mathcal{T}^{\text{id}}_{\tau\kappa} = \mathcal{T}^{\text{id}}_{\kappa\tau}$. If $\tau\kappa$ and $\kappa\tau$ are idempotent, that is, $\mathcal{T}^{\text{id}}_{\tau\kappa} = \mathcal{T}_{\tau\kappa}$, respectively $\mathcal{T}^{\text{id}}_{\kappa\tau} = \mathcal{T}_{\kappa\tau}$, then τ and κ are compatible, that is, $\tau\kappa = \kappa\tau$ and $Q_\tau Q_\kappa = Q_\kappa Q_\tau$.

Proof
We know this already; however, observe that we can obtain $\mathcal{T}^{\text{id}}_{\tau\kappa}$ as the union of $\mathcal{T}_{(\tau\kappa)\ldots(\tau\kappa)}$ for compositions of finitely many factors $\tau\kappa$, and similar for $\mathcal{T}^{\text{id}}_{\kappa\tau}$. Hence first $\mathcal{T}^{\text{id}}_{\tau\kappa} \supset \mathcal{T}_{\kappa\tau}$ follows and then also $\mathcal{T}^{\text{id}}_{\tau\kappa} \supset \mathcal{T}^{\text{id}}_{\kappa\tau}$. Symmetry in τ and κ then yields $\mathcal{T}^{\text{id}}_{\tau\kappa} = \mathcal{T}^{\text{id}}_{\kappa\tau}$. □

For $M <\in \underline{\mathcal{G}}$ let $j_{\tau\kappa} : M \to Q_\kappa Q_\tau(M)$ be the canonical morphism obtained as the composition $M \underset{j_\tau}{\longrightarrow} Q_\tau(M) \underset{j'_\kappa}{\longrightarrow} Q_\kappa Q_\tau(M)$, the latter j'_κ also being the localization morphism. Even if $\tau\kappa$ is not a hereditary torsion theory, we still have that $j_{\tau\kappa}(M) = 0$ if and only if $M \in \mathcal{T}_{\tau\kappa}$. On the other hand $j_{\tau\kappa}(M) = 0$ does not yield that $Q_\kappa Q_\tau(M) = 0$! But still $Q_\kappa Q_\tau(M)$ is in $\mathcal{T}_{\tau\kappa}$ because $Q_\tau(M)/\kappa Q_\tau(M)$ is τ-torsion since $\kappa Q_\tau(M)$ contains $M/\tau(M)$ and we have $Q_\kappa Q_\tau(M) \supset Q_\tau(M)/\kappa Q_\tau(M) \supset 0$, with $Q_\kappa Q_\tau(M)/(Q_\tau(M)/\kappa Q_\tau(M))$ being κ-torsion.

In finishing this section we now restrict attention to $\underline{\mathcal{G}} = A$-mod for some Noetherian ring A. To $\tau\kappa$ we associate a filter of left ideals $\mathcal{L}(\tau\kappa)$ in A, taking it to consist of those left ideals of A containing a left ideal $\sum'_\alpha I_\tau x^\alpha_\kappa$ where $I_\kappa = \sum'_\alpha A x^\alpha_\kappa$ is in $\mathcal{L}(\kappa)$ and $I_\tau \in \mathcal{L}(\tau)$, letting $\mathcal{L}(\tau)$ and $\mathcal{L}(\kappa)$ be the Gabriel filters of τ respectively κ, that is, $L \in \mathcal{L}(\tau)$ if and only if $A/L \in \mathcal{T}_\tau$.

Proposition 3.10
For $M \in A$-mod, the following statements are equivalent:

i. $j_{\tau\kappa}(M) = 0$.

ii. $M \in \mathcal{T}_{\tau\kappa}$.

iii. For every $m \in M$ there is an $L \in \mathcal{L}(\tau\kappa)$ such that $Lm = 0$.

Proof

Only the implication ii $\Leftrightarrow$ iii needs some explanation. If $M \in \mathcal{T}_{\tau\kappa}$ and $m \in M$, then, since $M/\tau M$ is κ-torsion, there is an $I_\kappa \in \mathcal{L}(\kappa)$ such that $I_\kappa m \subset \tau M$. Pick a finite set of generators $\{x_\kappa^\alpha, \alpha\}$ for I_κ and select I_τ^α in $\mathcal{L}(\tau)$ such that $I_\tau^\alpha x_\kappa^\alpha = 0$. Hence there is an $I_\tau \in \mathcal{L}(\tau)$ such that $(\sum'_\alpha I_\tau x_\kappa^\alpha)m = 0$, put $L = \sum'_\alpha I_\tau x_\kappa^\alpha$. Conversely assume iii. Take $m \in M$ and $L \in \mathcal{L}(\tau\kappa)$ such that $Lm = 0$, say $L = \sum'_\alpha I_\tau x_\kappa^\alpha$, $I_\tau \in \mathcal{L}(\tau)$, $I_\kappa = \sum_\alpha A x_\kappa^\alpha \in \mathcal{L}(\kappa)$. Then for every α, $I_\tau x_\kappa^\alpha m = 0$, hence $x_\kappa^\alpha m \in \tau(M)$. Therefore $A x_\kappa^\alpha m \subset \tau M$ or $I_\kappa m \subset \tau M$ and thus $M/\tau M$ is κ-torsion establishing ii.

□

If $\mathcal{L}(\tau)$ has a cofinal system consisting of two-sided ideals, τ is said to be **symmetric**. For specific rings, symmetric localization has been used a lot; for example, for fully bounded Noetherian rings or for rings satisfying polynomial identities (PI-rings) all localizations are symmetric.

Corollary 3.6

If τ is symmetric, then $\mathcal{L}(\tau\kappa)$ has a filter basis consisting of left ideals $I_\tau J_\kappa$ for $I_\tau \in \mathcal{L}(\tau)$, $J_\kappa \in \mathcal{L}(\kappa)$.

Using words in hereditary torsion theories and passing to corresponding filters $\mathcal{L}(\tau_1 \ldots \tau_d)$ with intersection of filters expressing the commutative union of such words (viewed as intersections in the noncommutative sense), we obtain an interpretation of the noncommutative topology of Lemma 3.1, T in the notation introduced in the paragraph preceding that lemma, in terms of filters of left ideals of A. This provides a rather concrete way of dealing with the noncommutative topology defined for a noncommutative algebra A completely expressed intrinsically in A-mod.

3.2 Affine Elements

In this short section we point out how the existence of a spectral Grothendieck representation allows us to consider affine elements in Top(A), $A \in \underline{\mathcal{R}}$, in turn allowing the definition of a Zariski topology of A.

Suppose, throughout this section, that Rep is a spectral GC representation as defined in Definition 3.3. For notational convenience, the functor Rep(f_γ) corresponding to a morphism $f_\gamma : A \to A(\gamma)$, associated to $\gamma \in$ Top(A), will be denoted by F_γ : Rep($A(\gamma)$) $\to$ Rep(A).

Definition 3.6

We say that $\gamma \in$ Top(A) is **Rep-affine** (or just affine once a GC representation has been fixed) if F_γ defines an isomorphism of Rep($A(\gamma)$) to i_γ (Rep(A), γ), where i_γ is the functor defined before Proposition 2.17, that is, the canonical inclusion functor of the quotient category with respect to γ in Rep(A).

The definition allows us to think of Q_γ : Rep(A) $\longrightarrow i_\gamma$(Rep(A), γ) as a functor from Rep(A) to Rep($A(\gamma)$), up to the isomorphism introduced by F_γ, since $i_\gamma a_\gamma = Q_\gamma$.

Proposition 3.11
The following properties hold for a Rep-affine $\gamma \in \text{Top}(A)$, $A \in \underline{\mathcal{R}}$:

a. *For every $M \in \text{Rep}(A(\gamma))$, $F_\gamma(M)$ is γ-torsion free.*

b. *For every $M \in \text{Rep}(A(\gamma))$, $F_\gamma(M)$ is γ-injective.*

c. *The localization functor Q_γ is exact and commutes with coproducts in* Rep(A).

d. *If $f_\gamma : A \to A(\gamma)$ is the morphism in $\underline{\mathcal{R}}$ corresponding to γ and $\tau \in \text{gen}(\gamma)$ in $\text{Top}(A)$, then we write $\widetilde{\tau}$ for $\widetilde{f_\gamma}(\tau)$, $\widetilde{f_\gamma} : \text{Top}(A) \to \text{Top}(A(\gamma))$ as in Proposition 3.2. Then $\widetilde{\gamma} = \xi_o(A(\gamma))$. For $X \in \text{Rep}(A(\gamma))$ we have: $F_\gamma Q_{\widetilde{\tau}}(X) = Q_\tau(F_\gamma(X))$. For $Y \in \text{Rep}(A)$ we have: $F_\gamma(Q_{\widetilde{\tau}}(G_\gamma Q_\gamma(Y))) = Q_\tau(Y)$. See proof of c for definition of G_γ.*

e. *$\text{Top}(A(\gamma))$ is lattice isomorphic to $\text{gen}(\gamma)$, $\mathcal{Q}_h(A(\gamma))$ is lattice isomorphic to $\text{gen}(\gamma)$ in $\mathcal{Q}_h(A)$.*

Proof
Proposition 3.11a and b follow immediately because $F_\gamma(M)$ is in $i_\gamma(\text{Rep.}(A), \gamma)$.

c. It is clear that $G_\gamma Q_\gamma$ is left adjoint to F_γ, where the functor

$$G_\gamma : i_\gamma(\text{Rep}(A), \gamma) \to \text{Rep}A(\gamma)$$

is defined by the isomorphism F_γ; that is, $G_\gamma F_\gamma$ is the identity of $\text{Rep}A(\gamma)$ and so forth. Therefore $G_\gamma Q_\gamma$ is an exact functor commuting with coproducts. Since $F_\gamma G_\gamma Q_\gamma = Q_\gamma$, the latter is also an exact functor commuting with coproducts.

d. By definition, an M in $\text{Rep}(A(\gamma))$ will be $\widetilde{\gamma}$-torsion exactly when $F_\gamma(M)$ is γ-torsion; however, by a we have that $F_\gamma(M)$ is γ-torsion free; hence $\widetilde{\gamma}$ will be the trivial torsion theory on $\text{Rep}(A(\gamma))$, which we denoted by $\xi_0(A(\gamma))$. If we establish for X in $\text{Rep}(A(\gamma))$ that $F_\gamma Q_{\widetilde{\tau}}(X) = Q_\tau(F_\gamma(X))$, then the second statement will follow by taking $X = G_\gamma Q_\gamma(Y)$ and observing that $Q_\tau(Q_\gamma(Y) = Q_\tau(Y)$ because $\tau \geq \gamma$.
For $X =_{A(\gamma)} A(\gamma)$ the statement above is actually ii in Proposition 3.2. For an arbitrary X in $\text{Rep}(A(\gamma))$, the exactness of F_γ allows us to reduce the proof of Proposition 3.2 considerably. Indeed $Q_{\widetilde{\tau}}(X)/X$ is $\widetilde{\tau}$-torsion; hence $F_\gamma Q_{\widetilde{x}}/F_\gamma(X)$ is τ-torsion; on the other hand, $Q_\tau(F_\gamma(X))/F_\gamma(X)$ is τ-torsion; hence $G_\gamma Q_\tau(F_\gamma(X))/G_\gamma F_\gamma(X)$ is $\widetilde{\tau}$-torsion. It follows that $G_\gamma Q_\tau(F_\gamma(X)) \subset Q_{\widetilde{\tau}}(X)$. Combined with the foregoing, this immediately yields that $F_\gamma Q_{\widetilde{\tau}}(X) = Q_\tau(F_\gamma(X))$.

e. If $\tau \in \text{gen}(\gamma)$, then τ is $(\text{Rep}(A), \gamma)$ compatible (or γ-compatible) as in Proposition 2.26. From d we derive that $G_\gamma i_\tau(\text{Rep}(A), \tau) = i_{\widetilde{\tau}}(\text{Rep}(A(\gamma)), \widetilde{\tau})$. It is easily verified that the correspondence $\widetilde{\tau} \to \tau$ defines a lattice isomorphism $\text{Top}A(\gamma) \to \text{gen}(\gamma) \subset \text{Top}(A)$. Since for this fact it is enough to establish such nice correspondence between the pretorsion classes of $\widetilde{\tau}$ and τ, respectively, it is enough to observe that $G_\gamma \tau(Y) = \widetilde{\tau} G_\gamma(Y)$ for Y

in Rep(A), $\tau(Y) = F_\gamma \widetilde{\tau} G_\gamma(Y)$. Since left exact preradicals τ (also $\widetilde{\tau}$) are determined by their pretorsion class, the result follows. □

If $\gamma \leq \tau$ are both affine, respectively γ and $\widetilde{\tau}$ for $\gamma \leq \tau$, then F_τ, respectively $F_\gamma F_{\widetilde{\tau}}$ define isomorphisms to the same category $i_\tau(\text{Rep}(A), \gamma)$, but from an isomorphism of Rep($A(\tau)$) and Rep($A(\gamma)(\widetilde{\tau})$) nothing can be concluded for $A(\tau)$ and $A(\gamma)(\widetilde{\tau})$.

Definition 3.7
A spectral Grothendieck representation Rep is **inductive schematic** if it is schematic in the sense of Definition 3.4, and for every $\gamma \leq \tau$ in Top(A), $A \in \underline{\mathcal{R}}$, the map f_{τ_γ} : $A(\gamma) \to A(\tau)$ coincides with $f_{\widetilde{\tau}} : A(\gamma) \to A(\gamma)(\widetilde{\tau})$, in particular $A(\tau) = A(\gamma)(\widetilde{\tau})$.

Corollary 3.7

1. *If Rep is inductive schematic, then for $\gamma \leq \tau$ in* Top(A) *there is equivalence between γ and τ being affine and γ and $\widetilde{\tau}$ being affine.*
2. *If P :* Top(A) $\to \underline{\mathcal{R}}$ *is a presheaf with values in $\underline{\mathcal{R}}$ over* Top$(A)^o$ *as in Proposition 3.3, then for $\gamma \in$* Top(A) *that is affine we have for every $\tau \in$* gen(γ) *that $A(\tau) = A(\gamma)(\widetilde{\tau})$). $P|$gen(γ) is a presheaf with values in $\underline{\mathcal{R}}$ associating to affine $\tau \in$ gen(γ) the object $P(A(\gamma)(\widetilde{\tau}))$.*

3.2.1 Observation and Example

When $\underline{\mathcal{R}}$ = $\underline{\text{Ring}}$ and Rep(A) is A-mod with the Fs being given by the usual restrictions of scalars with respect to ring morphisms, then this is an example of an inductive schematic Grothendieck representation. Other derived examples are obtained by looking at the category of graded rings and categories of graded left modules or by restricting further to positively graded rings and suitable quotient categories of graded left modules. The latter is the main example of a quotient GC representation, to be studied in the next section.

3.3 Quotient Representations

As before we consider the category $\underline{\mathcal{R}}$ and a Grothendieck representation Rep. A **topological nerve** for Rep is obtained by associating to each A in $\underline{\mathcal{R}}$ an $\eta_A \in$ Top(A) such that for any $f : A \to B$ in $\underline{\mathcal{R}}$ we have that: $\eta_B \leq \widetilde{f}(\eta_A)$ in Top(B). Since the latter domination relation is demanded with respect to any $f : A \to B$ in $\underline{\mathcal{R}}$, the notion of a topological nerve seems to be a rather narrow one; in concrete situations the choice of η_A in each Top(A) follows from radical-like constructions. Fix notation as follows:

$$i_A : (\text{Rep}(A), \eta_A) \to \text{Rep}(A)$$
$$a_A : \text{Rep}(A) \to (\text{Rep}(A), \eta_A)$$
$$Q_A : Q_{\eta_A} = i_A a_A$$

To a topological nerve $\underline{\eta}$ for Rep we associate a functor (Rep, $\underline{\eta}$) associating to A in $\underline{\mathcal{R}}$ the Grothendieck category (Rep, $\underline{\eta}$)(A) = (Rep(A), η_A), and to a morphism $f : A \to B$ in $\underline{\mathcal{R}}$ we associate the functor:

$$a_A F i_B : (\text{Rep}(B), \eta_B) \to (\text{Rep}(A), \eta_A), \text{ with } F = f^o.$$

When we consider a morphism $g : B \to C$ in $\underline{\mathcal{R}}$ we find that the composition:

$$(a_A F i_B)(a_B G i_C) : (\text{Rep}(C), \eta_C) \to \text{Rep}(A, \eta_A)$$

corresponds to the $\underline{\mathcal{R}}$-morphism $g \circ f$. So we have to establish that:

$$(a_A F i_B)(a_B G i_C) = a_A F G i_C;$$

this follows from the following more general fact.

Proposition 3.12
With notation as above: for any $f : A \to B$ in $\underline{\mathcal{R}}$ and M in Rep(B) *we have that $a_A F i_B a_B(M) = a_A F(M)$.*

Proof
Since $\eta_B \leq \tilde{f}(\eta_A)$ (we shall write $\tilde{\eta}_A$ for $\tilde{f}(\eta_A)$) and $Q_B(M)/M$ is η_B-torsion in Rep(B), it follows that it is $\tilde{\eta}_A$-torsion. By definition of $\tilde{\eta}_A$ we find that $F i_B a_B(M)/F(M)$ is η_A-torsion in Rep(A). From the exact sequence in Rep(B):

$$0 \longrightarrow \eta_B(M) \longrightarrow M \longrightarrow i_B a_B(M)$$

we obtain the following exact sequence in Rep(A):

$$0 \longrightarrow F\eta_B(M) \longrightarrow F(M) \longrightarrow F i_B a_B(M) \longrightarrow T \longrightarrow 0$$

where T is η_A-torsion in Rep(A) in view of the foregoing.

Since $\eta_B(M)$ is $\tilde{\eta}_A$-torsion, $F\eta_B(M)$ is η_A-torsion and therefore $Q_A(F(M)) = Q_A(F i_B a_B(M))$ or also $a_A F(M) = a_A F i_B a_B(M)$. □

Now we would like to conclude that (Rep, $\underline{\eta}$) is again a Grothendieck representation; however, since we have not assumed any finiteness condition on the η_A (the weakest workable one being that Q_A commutes with arbitrary direct sums or products), the appearance of Q_A in the definition may create a conflict with desirable properties like commutation with products. Since even in **Ring** some objects are Noetherian and some are not, we found it suitable to introduce a slight generalization of GC representation. A **generalized GC representation** is defined by weakening the definition of a GC representation such that for $f : S \to R$ in $\underline{\mathcal{R}}$, F : Rep(R) $\longrightarrow$ Rep(S) is only assumed to commute with finite products.

Corollary 3.8
With notation as above: to the composition $gf : A \to C$ we have associated $a_A F G i_C$, hence associating $f^{\square} = a_A F i_B$ to $f : A \to B$ in $\underline{\mathcal{R}}$ defines a generalized

GC representation (Rep, $\underline{\eta}$). *Observe in particular that* $a_A I_M i_A = I_M$ *for all* $M \in$ (Rep(A), η_A).

Proof

First, $a_A F G i_C = (a_A F i_B)(a_B G i_C)$ follows from the foregoing proposition, so it will suffice to verify that $a_A F i_B$ is exact and commutes with finite products and coproducts. This is clear except perhaps for the exactness because i_B need not be an exact functor. Consider an exact sequence in (Rep(B), η_B):

$$0 \longrightarrow M' \longrightarrow M \longrightarrow M'' \longrightarrow 0$$

Then we obtain an exact sequence in Rep(B),

$$0 \longrightarrow i_B M' \longrightarrow i_B M \longrightarrow i_B M'' \longrightarrow T_B \longrightarrow 0$$

where T_B is η_B-torsion in Rep(B).

Applying the exact functor F to the foregoing yields the following exact sequence in Rep(A):

$$0 \longrightarrow F i_B(M') \longrightarrow F i_B(M) \longrightarrow F i_B(M'') \longrightarrow F(T_B) \longrightarrow 0.$$

Since T_B is η_B-torsion it is certainly $\widetilde{\eta}_A$-torsion and thus $F(T_B)$ is η_A-torsion in Rep(A). Now we apply the exact functor a_A : Rep(A) $\rightarrow$ (Rep(A), η_A), and we arrive at an exact sequence in (Rep(A), η_A):

$$0 \longrightarrow a_A F i_B(M') \longrightarrow a_A F i_B(M) \longrightarrow a_A F i_B(M'') \longrightarrow 0.$$

The representation (Rep, $\underline{\eta}$) associated with a topological nerve $\underline{\eta}$ is called the **quotient generalized GC representation** of Rep.

If $\underline{\eta}^1$ and $\underline{\eta}^2$ are topological nerves, then $\underline{\eta}^1 \wedge \underline{\eta}^2$, defined by associating $\eta_A^1 \wedge \eta_A^2$ to A in $\underline{\mathcal{R}}$, is again a topological nerve. Note that this does not hold for $\vee$.

In the definition of spectral representation we have included the faithfulness of Rep. This was purely a matter of taste, and since the classical examples satisfy the condition it seems harmless. However, by passing to quotient categories we should perhaps not insist on faithfulness; therefore, we say that Rep is **weakly spectral** if it satisfies the conditions for a spectral representation, except that it need not be faithful.

Theorem 3.1

With conventions and notation as before:

i. *If Rep is measuring* $\underline{\mathcal{R}}$, *then so is* (Rep, $\underline{\eta}$).

ii. *If Rep is weakly spectral, then so is* (Rep, $\underline{\eta}$).

Proof

i. Assume that Rep is measuring $\underline{\mathcal{R}}$; hence there are ${}_A A$ in Rep(A) for A in $\underline{\mathcal{R}}$ with for $f : S \rightarrow R$ in $\underline{\mathcal{R}}$ a corresponding $f^* :_S S \rightarrow_S R = \text{Rep}(f)({}_R R)$ such that $I_S^* = I_{{}_S S}$, $(f \circ g)^* = \text{Rep}(g)(f^*) \circ g^*$ for $g : T \rightarrow S$ in $\underline{\mathcal{R}}$. For A

in $\mathcal{R}$ we consider $a_A({}_AA) \in \text{Rep}(A, \eta_A)$. To $f : A \to B$ in $\mathcal{R}$ we associate $f^{\square} = a_A f^o i_B$ and we also obtain $f^{\#} : a_A({}_AA) \to f^{\square}(a_B({}_BB))$. Taking Proposition 3.12 into account, we have:

$$f^{\square}(a_B({}_BB)) = a_A f^o({}_BB) = a_A({}_AB)$$

and for $f^{\#}$ we take $a_A(f^*)$, that is, the localized map of f^*. Obviously $(I_A)^{\#} = I_{a_A(A)}$; the composition rule follows from Corollary 3.8.

ii. For every $\gamma \in \text{Top}(A)$, A in $\mathcal{R}$, we have $A(\gamma)$ in $\mathcal{R}$ and $f_\gamma : A \to A(\gamma)$ in $\mathcal{R}$ such that $f_\gamma^* :_A A \to f_\gamma^o({}_{A(\gamma}A(\gamma))$ is exactly the localization morphism ${}_AA \to Q_\gamma({}_AA)$ in Rep(A). Consider some $\overline{\gamma} \in \text{Tors}(\text{Rep}(A), \eta_A)$; then $\overline{\gamma}$ corresponds to a $\gamma \in \text{Top}(A)$ such that $\gamma \geq \eta_A$. Since the functor a_A is exact and commutes with finite products, we may take γ to be given by its torsion class consisting of M in Rep(A) such that $a_A(M)$ is $\overline{\gamma}$-torsion in (Rep(A), η_A). We find that $a_A(f_\gamma^*) = f_\gamma^{\#}$ as follows:

$$\begin{array}{rcl} f_\gamma^{\#} : a_A({}_AA) & \to & a_A f_\gamma^o i_{A(\gamma)}({}_{A(\gamma)}A(\gamma)) \\ & & \| \\ & & a_A f_\gamma^o({}_{A(\gamma)}A(\gamma)) \\ & & \| \\ & & a_A(Q_\gamma({}_AA)) \end{array}$$

where the first equality follows from Proposition 3.12 and the second from the choice of $A(\gamma)$. Since $\gamma \geq \eta_A$ they are compatible kernel functors and we have $Q_{\eta_A} Q_\gamma({}_AA) = Q_\gamma({}_AA) = Q_\gamma Q_{\eta_A}({}_AA) = Q_{\overline{\gamma}} Q_{\eta_A}({}_AA)$. Therefore we arrive at $a_A(Q_\gamma({}_AA)) = a_{\overline{\gamma}} Q_{\eta_A}({}_AA) = a_{\overline{\gamma}}({}_AA)$ and $f_\gamma^{\#}$ is the localization map corresponding to $\overline{\gamma}$. □

Remark 3.1

1. The theory of GC representations extends to the case of generalized GC representations.
2. Given $f : A \to B$, then $a_A f^o i_B(M) = 0$ if and only if $i_B M$ is $\widetilde{\eta}_A$-torsion; that is, M is $\overline{\eta}_A$-torsion where $\overline{\eta}_A$ is induced on (Rep(B), η_B) by $\widetilde{\eta}_A$; thus even if Rep is faithful (Rep, $\underline{\eta}$) need not be.

It is possible to obtain the equivalent of Proposition 3.2 for quotient representations.

Proposition 3.13
Let Rep be a spectral GC representation; then (Rep, $\underline{\eta}$) *is a weakly spectral generalized GC representation such that for* $\tau \geq \gamma \geq \eta_A$ *in* Top(A)*:*

i. $Q_{\overline{\eta}_A}(f_{\tau_\gamma}^{\square}({}_{A(\tau)}A(\tau))) = Q_{\widetilde{\tau}}({}_{A(\gamma)}A(\gamma))$

ii. $f_\gamma^{\square}(a_{\widetilde{\tau}}({}_{A(\gamma)}A(\gamma))) = Q_\tau({}_AA) = Q_{\widetilde{\tau}}(Q_A({}_AA))$

Proof

i. To $\tau \geq \gamma \geq \eta_A$ in Top(A) we have associated $\widetilde{f}_\gamma(\tau) \geq \widetilde{f}_\gamma(\gamma) \geq \widetilde{f}(\eta_A) \geq \eta_{A(\gamma)}$ as well as $\overline{\tau} \geq \overline{\gamma}$ in Tors(Rep(A), η_A). Now $f^{\square}(M) = 0$ exactly if M is $\widetilde{\eta}_A$-torsion. Moreover, an object that is $\widetilde{\gamma} = \widetilde{f}_\gamma(\gamma)$-torsion is certainly $\widetilde{\tau} = \widetilde{f}_\gamma(\tau)$-torsion. Now one may follow the lines of the proof of Proposition 3.2, defining $M_\tau(\gamma)$ as $f^{\square}_{\tau_\gamma}(_{A(\tau)}A(\tau))$ and replacing $f^o_\gamma = \text{Rep}(f_\gamma)$ by $f^{\square}_\gamma$, f^*_τ by $f^{\#}_\tau$, and so forth. It follows that $f^{\square}(M_\tau(\gamma))$ is $\widetilde{\eta}_A$-torsion free and $M_\tau(\gamma)$ is $\eta\eta_A$-torsion free. Thus $f^{\square}_\gamma(\widetilde{\tau}M_\tau(\gamma)) = 0$ leads to $\widetilde{\tau}M_\tau(\gamma)$ being $\widetilde{\eta}_A$-torsion contradicting the foregoing unless $\widetilde{\tau}(M_\tau(\gamma)) = 0$. As in the proof of Proposition 3.2 we then arrive at

$$f^{\square}_\gamma(Q_{\widetilde{\tau}}(_{A(\gamma)}))/M_\tau(\gamma) = 0;$$

hence $Q_{\widetilde{\tau}}(_{A(\gamma)}A(\gamma))$ is $\widetilde{\eta}_A$-torsion over $M_\tau(\gamma)$ and i follows.

ii. We have established earlier that (Rep, $\underline{\eta}$) is weakly spectral; hence we have:

$$f^{\square}_\gamma f^{\square}_{\tau_\gamma}(_{A(\tau)}A(\tau)) = Q_{\widetilde{\tau}}(Q_{\eta_A}(_A A)),$$

Hence $f^{\square}_\gamma(M_\tau(\gamma))$ is $\widetilde{\tau}$-closed and consequently $f^{\square}_\gamma(Q_{\widetilde{\tau}}(_{A(\gamma)}A(\gamma))) = f^{\square}_\gamma M_\tau(\gamma)$ follows; again it is clear that ii. follows from the latter. □

3.3.1 Project: Geometrically Graded Rings

Consider a $\mathbb{Z}$-graded Noetherian ring R, $R = \oplus_{n\in\mathbb{Z}} R_n$; that is, R_0 is a Noetherian subring of R. The subset $\delta = \sum_{n\neq 0} R_{-n}R_n$ is an ideal of R_0 and $I = \delta \oplus (\oplus_{n\neq 0} R_n)$ is a graded ideal of R such that $I_0 = \delta$. We say that an ideal J of R_0 is **invariant** if for all $n \in \mathbb{Z}$, $R_n J R_{-n} \subset J$. Clearly δ is an invariant ideal. In general, an ideal J of R_0 is invariant if and only if $J \oplus (\oplus_{n\neq 0} R_n)$ is an ideal of R. Consequently, every ideal J of R_0 such that $J \supset \delta$ is invariant. A torsion theory κ given by its filter $\mathcal{L}(\kappa)$ is said to be invariant if $J \in \mathcal{L}(\kappa)$ entails $\sum_{n\neq 0} R_n J R_{-n} \in \mathcal{L}(\kappa)$. For example, the torsion theory with filter $\mathcal{L}(\delta) = \{L$ left ideal of R_0, $L \supset J$ where J is an ideal of R_0 such that $\delta \subset \text{rad}(J)\}$ is invariant. Indeed, if $L \in L(\delta)$ and J is as above, look at $L' = \sum_{n\neq 0} R_n J R_{-n}$; the Noetherian property yields that $\text{rad}(J)^m \subset J$; hence $(R_n \text{rad}(J) R_{-n})^m \subset R_n \text{rad}(J)^m R_{-m}$ for all $n \in \mathbb{Z}$; thus if some prime ideal P of R_0 contains J', then it contains $\sum_{n\neq 0} R_n \text{rad}(J) R_{-n}$ and a fortiori: $P \supset \sum_{n\neq 0} R_n \delta R_{-n}$. Then from $P \supset R_n R_{-n} R_n R_{-n}$ for all $n \neq 0$, $P \supset R_n R_{-n}$, and the claim is clear. It would already be an acceptable generalization of projective theory to be able to deal with the case where R_0 is in the center of R, or more generally whenever R is a centralizing extension of R_0; however, a far less restrictive condition allows us to develop the theory we are looking for. The graded ring R is said to be **o-normal** if for all $n \in \mathbb{Z}$ we have $R_{-n}(R_n R_{-n}) \subset (R_n R_{-n})R_{-n}$, R is **geometrically graded** if R is Noetherian, R_0 is central in R, and R is generated over R_0 by $R_1 \cup R_{-1}$ as a ring extension.

Proposition 3.14
Let R be o-normal as before. If $\delta = R_o$, then there exists a $d \in \mathbb{Z}$ such that $R_d R_{-d} = R_o = R_{-d} R_d$.

Proof
Fix $a \in \mathbb{Z}$, $a \neq 0$, and look at $(R_a R_{-a})^n$ for $n \in \mathbb{N}$.

For $n \neq 0 : (R_a R_{-a})^n = (R_a R_{-a})(R_a R_{-a}(\ldots) \subset R_a(R_a R_{-a})R_{-a}(R_a R_{-a})\ldots$

The latter equals $R_a^2 R_{-a}^2(R_a R_{-a})\ldots$. Repeating the passing of $R_a R_{-a}$ over some R_{-a} to the left, we arrive at $(R_a R_{-a})^n \subset R_a^n R_{-a}^n \subset R_{na} R_{-na}$, or $R_a R_{-a} \subset \text{rad}(R_{na} R_{-na})$. Now observe the following:

$$(R_a R_{-a})(R_{-a} R_a) = R_{-a}(R_a R_{-a})R_a \subset (R_a R_{-a})(R_{-a} R_a)$$

and the latter is contained in $R_a R_{-a}$ since that is an ideal of R_0. Thus $R_{-a} R_a \subset \text{rad}(R_a R_{-a})$. If $1 \in \delta$ there exist finitely many $a_1, a_2, \ldots, a_m \in \mathbb{Z}$ such that $1 \in \sum_{i=1}^m R_{a_i} R_{-a_i}(*)$. Since $R_{-a} R_a \subset \text{rad}(R_a R_{-a})$ for all $a \in \mathbb{Z}$, it follows that a_i and $-a_i$ may be interchanged in the expression (*) and this leads to $1 \in \sum_{i=1}^m R_{a_i} R_{-a_i}$ (up to renaming the a_i) where now $a_i \in \mathbb{N}$ for all $i = 1, \ldots, m$. Now from $R_a R_{-a} \subset \text{rad}(R_{na} R_{-na})$ for all $n \in \mathbb{N}$, we obtain $\text{rad}(\sum_{i=1}^m R_{a_i} R_{-a_i}) \subset \text{rad}(R_d R_{-d})$, where $d \in \mathbb{N}$ is the lowest common multiple of $a_1, \ldots, a_m \in \mathbb{N}$. Consequently, $R_o = \text{rad}(R_d R_{-d})$ or $R_o = R_d R_{-d}$. For $R_{-d} R_d$ calculate, using o-normality applied to $-d$ $R_{-d} R_d = R_{-d}(R_d R_{-d})R_d \subset R_d(R_{-d} R_d)R_d$; thus $R_d(R_{-d} R_d)R_{-d} \subset R_d(R_{-d} R_{-d} R_d R_d)R_{-d}$. Now, substituting $R_d R_{-d} = R_0$ leads to $R_o \subset R_{-d} R_d$. □

The dth**-Veronese subring** of R is the $\mathbb{Z}$-graded ring $R(d)$ defined by putting $R(d)_n = R_{nd}$. We say that R is d-strongly graded if $R(d)$ is strongly graded; that is, $R(d)_1 R(d)_{-1} = R(d)_{-1} R(d)_1 = R(d)_0$. As a corollary to the foregoing proposition, it is clear that $\delta = R_o$ means exactly that R is d-strongly graded for some suitable $d \in \mathbb{N}$. For **"rigid" torsion** theories see also Section 3.4.

Proposition 3.15
If R is a o-normal (left) Noetherian $\mathbb{Z}$-graded ring and τ is a perfect rigid torsion theory on R-gr with graded filter $\mathcal{L}^g(\tau)$ then, if $\delta \neq 0$ and $R\delta \subset \mathcal{L}^g(\tau)$, $Q_\tau^g R$ is d-strongly graded for some $d \in \mathbb{N}$.

Proof
Assumptions imply that $\text{rad}(\delta)$ is finitely generated as a left ideal, so $\text{rad}(\delta) = \text{rad}(\sum_{i=1}^m R_{a_i} R_{-a_i})$ with $a_1, \ldots, a_m \in \mathbb{N}$ (argumentation as in the foregoing proof). Hence, $\text{rad}(\delta) = \text{rad}(R_d R_{-d})$, $d = l.c.m(a_1, \ldots, a_m)$. Since τ is perfect and $R\delta \in \mathcal{L}^g(\tau)$, we have $S\delta = S$ where $S = Q_\tau^g(R)$. Taking parts of degree zero: $S_0\delta = S_0$ and thus $S_0\text{rad}(R_d R_{-d}) = S_0$ as well as $S_0\text{rad}(R_d R_{-d})^m = S_0$ for any $m \in \mathbb{N}$. Since for some $m_0 \in \mathbb{N}$ we have: $(\text{rad}(R_d R_{-d}))^{m_0} \subset R_d R_{-d}$ (R_0 is Noetherian!) it also follows that $S_0 R_d R_{-d} = S_0$. The obvious inclusions $S_d \supset R_d$, $S_{-d} \supset R_{-d}$ then yield $S_d S_{-d} = S_0$. Again, applying the foregoing argumentation to $R_{-d} R_d$ and noting that $\text{rad}(R_d R_{-d}) = \text{rad}(R_{-d} R_d)$ leads to $S_d S_{-d} = S_{-d} S_d = S_0$ as desired. □

Exercise 3.1
Is S o-normal when R is o-normal ? The foregoing results reduce the problem of studying the noncommutative geometry (of projective type) to the d-strongly graded situation by passing to a suitable open subset corresponding to localization at $\mathcal{L}(\delta)$. However, the case $\delta = 0$ is excluded there. As an exercise, prove the following lemma.

Lemma 3.4
In case $\delta = 0$ but R is a domain, R is necessarily positive or negatively graded.

Let us return to the philosophy of [49], that is, view noncommutative geometry of associative (schematic) algebras via the localizations at Ore sets. For a homogeneous Ore set T in R, $T(d) = T \cap R(d)$ is an Ore set in $R(d)$.

Indeed, for $t \in T(d), r \in R(d)$ we find $t' \in T, r' \in R$, such that $t'r = r't$. Hence $(t')^d r = (t')^{d-1} r't$ with $(t')^d \in T(d)$ and $(t')^{d-1}r' \in R(d)$. But we are not so lucky that every Ore set of $R(d)$ is of the form $T(d)$! This problem may be circumvented by developing a noncommutative "weighted" space generalizing the commutaive *weighted projective spaces*; we leave this as a project of independent interest. For geometrically graded rings, all methods developed earlier work perfectly.

Proposition 3.16
Let R be geometrically graded and τ a perfect rigid torsion theory on R-gr given by its graded filter $\mathcal{L}^g(\tau)$.

i. *If $\delta \neq 0$ and $R\delta \in \mathcal{L}^g(\tau)$, then $S = Q^g_\tau(R)$ is strongly graded.*

ii. *If $\delta = 0$ and $I \in \mathcal{L}^g(\tau)$, then $S_0 = S_{-1}S_1 = S_{-n}S_n$ for $n \geq 0$.*

In case τ corresponds to an Ore set T of R that is homogeneous and not contained in R_0, then S is strongly graded.

Proof
An $r \in R$ can be written as a sum of monomials of type $r_0x_1 \ldots x_n$ with $r_0 \in R_0$, x_i in $R_1 \cup R_{-1}$. In case $x_i \in R_1$ and $x_{i+1} \in R_{-1}$ or conversely, then $x_ix_{i+1} \in R_0$ and thus central in R. Consequently, such a monomial is in $R_1^d R_{-1}^e$ or in $R_{-1}^e R_1^d$ for suitable e and d in $\mathbb{N}$. In a similar way we verify that $R_nR_1^n$, $R_m = R_m = R_{-1}^{-m}$ for $n \geq 0$, $m \leq 0$.

i. If $\delta \neq 0$, then for some $d \in \mathbb{N}$, $S_0R_dR_{-d}R_d = S_0$; this follows from argumentation as in Proposition 3.15. Now $R_dR_{-d} = R_1^dR_{-1}^d = (R_1R_{-1})^d$ follows, and from $S_o(R_1R_{-1})^d = S_0$ it follows that $S_o(R_1R_{-1}) = S_0$. In a similar way, $S_0(R_{-1}R_1) = S_0$ follows from $S_0(R_{-d}R_d) = S_0$. This leads to $S_1S_{-1} = S_0 = S_{-1}S_1$.

ii. $\delta = 0$ if and only if $R_1R_{-1} = R_{-1}R_1 = 0$ or $R_nR_m = 0$ for $n > 0, m < 0$; thus $\delta = 0$ if and only if R is either positively or negatively graded depending upon whether $R_1 \neq 0$ or $R_{-1} \neq 0$. We deal with the positively graded case; the other case is similar. Put $R = \oplus_{n \geq 0} R_n$, $I = \oplus_{n>0} R_n = R_+$. The assumption $I \in \mathcal{L}^g(\tau)$, τ being perfect, yields $S = SI$. Hence $S_0 = \sum_{n>0} S_{-n}T_n = \sum_{n>0} S_{-n}R_n$. Look at $s_{-n}r_n$ with $s_{-n} \in S_{-n}$, $R_n \in R_n$. For some $L \in \mathcal{L}^g(\tau)$ we have that

$$(*) \qquad L_ps_{-n}r_n \supset R_{p-n}R_n \subset R_p$$

thus for any $p > 0$:

$$S_{-p}L_ps_{-n}r_n \subset S_{-p}T_p = S_{-p}R_1^{p-1}R_1 \subset S_{-1}R_1 \subset S_{-1}S_1$$

Observe that $\sum_{p>0} S_{-p}L_p = (SL)_0 = S_0$ (as $L \subset \mathcal{L}^g(\tau)$); consequently, $S_0 s_{-n} r_n \subset S_{-1}S_1$ with n arbitrary. From $S_0 = \sum_{n>0} S_{-n}R_n$ we obtain $S_0 = S_{-1}S_1$. Note that in (*) $L_0 = 0$ because if $L \in \mathcal{L}^g(\tau)$, then $I \cap L \in \mathcal{L}^g(\tau)$ with $(I \cap L)_0 = 0$, so we may replace L by $L \cap I$ without loss of generality. Observe that the foregoing does not imply that also $S_1 S_{-1} = S_0$! However, when τ is associated with a homogeneous nontrivial (left) Ore set T, then for $y \in S_0$ look at yt_m with $t_m \in T \cap R_m$ for $m > 0$. Since t_m is invertible in S with $t_m^{-1} \in S_{-m}$ we may look at $(yt_m)t_m^{-1} = y \in S_m S_{-m}$. Now from $S_{-1}S_1 = S_0$ we derive that $S_m = S_1^m$ (because $S_m = S_m S_0 = S_m S_{-1} S_1$, thus $S_m = S_{m-1}S_1$ and by repetition of this argumentation we obtain $S_m = S_1^m$). Finally we obtain $y \in S_1^m S_{-m} = S_1(S_1^{m-1}S_{-m}) \subset S_1 S_{-1}$ and consequently $S_0 = S_1 S_{-1}$ as desired. □

To the graded ring R we associate a rigid torsion theory κ_R defined by its graded filter $\mathcal{L}^g(\kappa_R)$, which is the graded filter (Gabriel topology) generated by $R\delta$ and I. Observe that $I = \delta \oplus (\oplus_{n\neq 0} R_n)$ is automatically in $\mathcal{L}^g(\kappa_R)$ because it contains $R\delta$ when $\delta \neq 0$. The situation of geometrically graded rings R with rigid torsion theories κ_R is also interesting because it provides us with a new example of a topological nerve.

Lemma 3.5
Let R and S be $\mathbb{Z}$-graded rings such that δ_R and δ_S are both either nonzero or both zero. If $f : R \to S$ is a morphism of graded rings, then $\kappa_S \leq \widetilde{f}(\kappa_R)$.

Proof
Since $f(R_n) \subset S_n$ for every $n \in \mathbb{Z}$, it is clear that $f(\delta_R) \subset \delta_S$. By definition $L \in \mathcal{L}(\widetilde{f}(\kappa_R))$ means that S/L is κ_R-torsion as an R-module; that is, L contains some L', $L' \in \mathcal{L}(\kappa_R)$. □

Consider the category $\underline{\mathcal{B}}$ of $\mathbb{Z}$-graded rings R with $\delta_R \neq 0$, taking just graded ring morphisms for the morphisms. Associating R-gr to R defines a Grothendieck representation such that $\{\kappa_R, R \in \underline{\mathcal{B}}\}$ is ε nerve. Therefore we may consider the quotient Grothendieck representation with respect to the nerve $\{\kappa_R, R \in \underline{\mathcal{B}}\}$.

Exercise 3.2
Develop the noncommutative geometry of "Proj$R$", which is defined by the noncommutative topology of the quotient category (R-gr, κ_R), together with the corresponding sheaf theory. Define **schematically graded** *rings as the class of graded (Noetherian) rings R such that there exists a finite set of homogeneous Ore sets $T_1, \ldots, T_n$ such that $\kappa_R = \kappa_{T_1} \wedge \ldots \wedge \kappa_{T_m}$; suitable generalizations may be defined by replacing the κ_{T_i} by perfect rigid torsion theories not necessarily stemming from Ore sets. In this case* Proj$R$ *defined on (R-gr, κ_R) satisfies all properties valid in the positively graded case. In particular, the schematic condition entails the existence of an affine cover (invoking Proposition 3.16). It is possible to establish a proof of Serre's global section theorem for* Proj$R$. *A new ingredient in this project consists in the study of the relation between the affine geometry of R_0, in terms of* SpecR_0 *say, and the projective geometry in*

terms of Proj$R$*, that is, (R-gr, κ_R). A subproject of this consists of a concrete algebraic approach when R is a ring satisfying polynomial identities (in particular when R is a finite module over its center) where a relation with the theory of maximal (R_0-)orders has to be investigated. More concretely, study the geometry when R is an R_0-order, R_0 is a Noetherian integrally closed domain of dimension n and δ defines a closed subvariety of* SpecR_0 *of dimension $n_1 < n$. Even more concrete, $n = 1$ and $n_0 = 0$, or $n = 2$ and $n_1 = 1$. There are new phenomena here when compared to the theory started in [47] or in L. Le Bruyn, M. Van den Bergh, and F. Van Oystaeyen, Graded Orders, Birkhauser Monographs (xxxx).*

3.4 Noncommutative Projective Space

As an example of the quotient representations introduced in the foregoing section we point out how the construction of projective spaces fits in that theory. For the category $\underline{\mathcal{R}}$ we now restrict attention to the category of positively graded k-algebras with graded k-algebra morphisms of degree zero for the morphisms; recall that a graded k-algebra is said to be connected if its part of degree zero is k; that is, A is a graded connected k-algebra if $A = k \oplus A_1 \oplus A_2 \oplus \ldots$. For geometry-oriented purposes we restrict attention to finite gradations in the sense that each A_n is a finite dimensional k-space and A is generated as a k-algebra by A_1. In that case, the positive part $A_+ = A_1 \oplus A_2 \oplus \ldots$ is finitely generated as a left (or right) ideal of A and moreover $AA_1 = A_+$ and the powers A_+^m form the basis of a Gabriel topology of a torsion theory that we denote by κ_+ and we write $\mathcal{L}(\kappa_+)$ for the Gabriel topology. Any $\tau \in A$-tors is said to be **rigid** if the graded torsion class T_τ^g is shift invariant; that is, if a graded A-module M is τ-torsion, then for every $n \in \mathbb{Z} : T(n)M$ is also τ-torsion, and conversely. The set of graded left ideals in $\mathcal{L}(\tau)$ is denoted by $\mathcal{L}^g(\tau)$; it is called the graded filter or Gabriel topology. Now if we start with τ, even one such that $\tau(M)$ is a graded submodule of M whenever M is graded, then τ need not be characterized by $\mathcal{L}^g(\tau)$. This is due to the fact that A need not be a generator for A-gr. On the positive side, if τ is rigid, then it is characterized by $\mathcal{L}^g(\tau)$. In any case κ_+ is a rigid torsion theory, so it is completely determined by the graded Gabriel topology $\mathcal{L}^g(\kappa_+)$.

For an arbitrary graded ring R we denote by R-rig the sublattices of R-tors of rigid graded torsion theories.

Let us write $\underline{\mathcal{R}}_k^g$ instead of $\underline{\mathcal{R}}$ in this section, in order to reflect the graded character and to fix the field k. If $g : R \to S$ is a morphism in $\underline{\mathcal{R}}_k^g$, then $g(R_+) \subset S_+$. Let us write $\kappa_+(R)$ respectively $\kappa_+(S)$ for the rigid graded torsion theory in R-tors, respectively S-tors, associated to R_+, respectively S_+. Obviously, S/S_+ is $\kappa_+(R)$-torsion, so it follows easily that $\kappa_+(S) \leq \tilde{g}(\kappa_+(R))$ where $\tilde{g} : R$-tors$\to S$-tors corresponds to g.

Lemma 3.6
The restriction of $\tilde{g}$ to R-rig defines $\tilde{g} : R$-rig$\to S$-rig.

Proof

Take $\tau \in R$-rig and look at $\widetilde{g}(\tau)$. If N is a $\widetilde{g}(\tau)$-torsion graded S-module, then ${}_RN$ is τ-torsion and every $T(m)_RN$ is then τ-torsion because τ is rigid. Now it is clear that ${}_R(T(m)N) = T(m)_RN$; thus $T(m)N$ is $\widetilde{g}(\tau)$-torsion for every $m \in \mathbb{Z}$ or $\widetilde{g}(\tau)$ is rigid. □

For a k-algebra A graded as before, we let Proj(A) be the Grothendieck category obtained as the Serre quotient category of finitely generated graded A-modules modulo graded A-modules of finite length, that is, if A-gr$_f$ denotes the category of finitely generated graded A-modules, then the localizing functor A-gr$_f$ → Proj(A) corresponds to the torsion class of the κ_+-torsion objects, which in this case are finite dimensional over k. The functor A-gr$_f$ → Proj(A) defines a lattice morphism:

$$\mathrm{Tors}^g(\mathrm{Proj}(A)) \to \mathrm{Tors}^g(A-\mathrm{gr}_f)$$

where $\mathrm{Tors}^g(-)$ stands for the lattice of graded torsion theories on the category specified.

The latter morphism restricts to Rig(Proj(A)) → Rig($A-$gr$_f$), where Rig(−) stands for the lattice of rigid graded torsion theories. For full detail on graded localization theory we refer the reader to C. Năstăsescu and F. Van Oystaeyen, *Graded Rings and Modules*, LNM 758, Springer Verlag [31], or Graded Ring Theory, North Holland.

In this section we restrict attention to the *commutative shadow*; that is, we deal with the torsion theories and leave the extension to graded radicals and noncommutative topology to the reader (this is a fairly straightforward graded version of the arguments of part of Section 3.1, after Proposition 3.3).

Lemma 3.7

With notation as before we have:

1. Rig($A-$gr$_f$) = *A-rig.*
2. Rig(Proj(A)) = $\mathrm{gen}_{\mathrm{rig}}(\kappa_+)$, *the latter denoting the set of rigid graded torsion theories* τ *in A-tors such that* $\tau \geq \kappa_+$, *that is,* $\mathrm{gen}_{\mathrm{rig}}(\kappa_+) = \mathrm{gen}(\kappa_+) \cap$ *A-rig.*

Proof

1. For any torsion theory on modules it is true that a module is torsion if and only if every finitely generated submodule is torsion. Hence the restriction of a torsion class to A-gr$_f$ does determine the torsion class in A-gr$_f$. Rigidity of the (graded) torsion class in A-gr is obviously equivalent to the rigidity of the corresponding torsion class in A-gr$_f$. From foregoing observations it follows easily that Rig(A-gr$_f$) = A-rig.
2. The map Rig(Proj(A)) → A-rig associates to a rigid (graded) torsion theory on the quotient category Proj(A) of A-gr$_f$, the rigid torsion theory it induces on A-gr$_f$, and thus on A-gr, that is, an element of gen(κ_+)∩A-rig. Conversely, that any $\tau \in \mathrm{gen}_{\mathrm{rig}}(\kappa_+)$ induces a rigid torsion theory on Proj(A) follows by checking

the transfer of rigidity; the bijective correspondence Tors(Proj(A)) $=$ gen(κ_+) follows from earlier observations (this also follows from Proposition 2.26) □

The lattice $\text{gen}_{\text{rig}}(\kappa_+)$, or in fact the category corresponding to it in the usual way, may be viewed as the projective version of Top(A) introduced in the ungraded situation as Tors(Rep(A)). So it makes sense to write $\text{Top}_{\text{proj}}(A) = \text{gen}_{\text{rig}}(\kappa_+)$. If $g : R \to S$ is in $\underline{\mathcal{R}}_\kappa^g$, then we have already introduced $\widetilde{g}$: R-rig$\to$ S-rig in Lemma 3.6, which may be viewed as a functor when the lattices are considered as categories in the usual way. However, there is a problem in constructing an associated map:

$$\text{Top}_{\text{proj}}(R) \longrightarrow \text{Top}_{\text{proj}}(S).$$

Indeed $g : R \to S$ does not necessarily define a "restriction of scalars" functor Proj(S)$\longrightarrow$ Proj(R)! On the positive side we have $\widetilde{g}(\kappa_+(R))$ in S-rig such that $\kappa_+(S) \leq \widetilde{g}(\kappa_+(R))$; therefore, the quotient category (S-gr_f, $\widetilde{g}(\kappa_+(R))$) is also a quotient category of Proj(S) such that Rig(S-gr_f, $\widetilde{g}(\kappa_+(R))$) may be identified to $\text{gen}_{\text{rig}}(\widetilde{g}(\kappa_+(R)))$ in S-rig via the map associated to the localizing functor S-gr_f $\longrightarrow$(S-gr_f, $\widetilde{g}(\kappa_+(R))$). So we may conclude that we obtain a functor Proj(S), $\widetilde{g}(\kappa_+(R))$ $\longrightarrow$ Proj(R), induced by the restriction of scalars with respect to g. By transitivity of the localization functors associated to $\kappa_+(S) \leq \widetilde{g}(\kappa_+(R))$ we actually find that:

$$(\text{Proj}(S), \widetilde{g}(\kappa_+(R))) = (S\text{--gr}_f, \widetilde{g}(\kappa_+(R)))$$

(or by the compatibility property deriving from $\kappa_+(S) \leq \widetilde{g}(\kappa_+(R))$).

In any case, the functor (Proj(S), $\widetilde{g}(\kappa_+(R))$) $\longrightarrow$ Proj(R) induces a lattice morphism:

$$\text{Rig}(\text{Proj}(R))) \longrightarrow \text{Rig}(S\text{--gr}_f, \widetilde{g}(\kappa_+(R))).$$

Observing that Rig(Proj(R)) $= \text{gen}_{\text{rig}}(\kappa_+(R))$ in R-rig, Rig(S-gr_f, $\widetilde{g}(\kappa_+(R))$) $= \text{gen}_{\text{rig}}(\widetilde{g}(\kappa_+(R)))$ in S-rig, we may conclude that the morphism g gives rise to a lattice morphism:

$$\text{gen}_{\text{rig}}(\kappa_+(R)) \longrightarrow \text{gen}_{\text{rig}}(\widetilde{g}(\kappa_+(R))) \hookrightarrow \text{gen}(\kappa_+(S))$$

defining a lattice morphism

$$\text{Top}_{\text{proj}}(R) \longrightarrow \text{gen}(\widetilde{g}(\kappa_+(R))) \hookrightarrow \text{Top}_{\text{proj}}(S).$$

In other words, the lattice morphism that we obtain here is obtained from g but not from a functor Proj(S) $\to$ Proj(R).

The above phenomenon is also present in the commutative scheme theory. It expresses the fact that, even when suitable localizations do carry over from R to S via $\widetilde{g}$, the scheme theory has to take into account that the underlying topological morphism can only be defined on an open subset of Proj(S) in fact given by gen($\widetilde{g}(\kappa_+(R))$) (viewed in the opposite lattice). □

The foregoing establishes the following proposition.

Proposition 3.17
To a morphism $g : R \to S$ in $\underline{\mathcal{R}}_k^g$ there corresponds a functor (deriving from a lattice morphism) $\mathrm{Top}_{\mathrm{proj}}(R) \longrightarrow \mathrm{Top}_{\mathrm{proj}}(S)$.

An arbitrary $g : R \to S$ does not allow us to relate finitely generated (graded) S-modules to finitely generated (graded) R-modules when S itself is not even finitely generated as an R-module. This makes it more natural to consider $(A\text{-}gr, \kappa_+(A))$ for any A in $\underline{\mathcal{R}}_k^g$, that is, without restricting to $A\text{-}g_f$. Let us write $\mathrm{PROJ}(A)$ *for the latter quotient category. In sheaf theoretical language this would mean that we focus on quasi-coherent sheaves rather than on coherent sheaves. The torsion objects with respect to $\kappa_+(A)$ in A-gr need not have finite length, but this does not affect any of the statements and results derived earlier.*

In the language of Section 3.3, where we put $\underline{\mathcal{R}} = \underline{\mathcal{R}}_k^g$, now we consider a Grothendieck representation associating A-gr (or $A\text{-}gr_f$) to A in $\underline{\mathcal{R}}_k^g$. A topological nerve κ_+ can now be obtained by letting n_A (as in Section 3.3) be $\kappa_+(A)$ as defined earlier in this section. For a morphism $g : A \to B$ in $\underline{\mathcal{R}}_k^g$ we do have that $\kappa_+(B) \leq \tilde{g}(\kappa_+(A))$ and so we arrive at a generalized Grothendieck representation; $(A\text{-}gr, \kappa_+(A))$ is then associated to A in $\underline{\mathcal{R}}_k^g$, which is the quotient generalized Grothendieck representation of Section 3.3 and in particular, from Theorem 3.1 it follows that it is measuring and weakly spectral; moreover, it satisfies the statements of Proposition 3.13. With notation as introduced in this section, the GC representation gr, respectively gr_f, *associating A-gr, respectively $A\text{-}gr_f$, to A in $\underline{\mathcal{R}}_k^g$, allows the quotient representation PROJ, respectively* Proj, *associating* $\mathrm{PROJ}(A)$, *respectively* $\mathrm{Proj}(A)$ *to A in $\underline{\mathcal{R}}_k^g$.*

3.4.1 Project: Extended Theory for Gabriel Dimension

In Section 2.6 we established how localization functors or torsion theories appear as a major example of noncommutative topology. In fact in view of the constructed scheme theory for schematic algebras (cf. [49]), the latter example has been the main motivation for the introduction of noncommutative topology in a more axiomatic way. Now unlike the Krull dimension, the Gabriel dimension is defined exactly in terms of torsion theories, so it is a possible instrument for calculating certain dimensions of noncommutative algebras, topologies, or other categorical structures. Let us recall some basic facts along the way to describing some possible projects.

The name **Gabriel dimension** is attributed by Gordon, Robson to a notion introduced by P. Gabriel in his thesis, [10], but here termed the Krull dimension. Since several notions of generalized Krull dimension became available later, the different names were used; in the book *Dimensions of Ring Theory* [33], the notion of Gabriel dimension is given for an arbitrary modular lattice. A first project could be to generalize this to noncommutative topologies or virtual topologies. We do not go into this, but turn to Grothendieck categories instead.

Let $\underline{\mathcal{G}}$ be a Grothendieck category. An object M of $\underline{\mathcal{G}}$ is said to be semi-Artinian if for every subobject M' of M such that $M \neq M'$ there exists a simple subobject in M/M'. The full subcategory of $\underline{\mathcal{G}}$ consisting of all semi-Artinian objects is easily seen to be a localizing subcategory, in other words to determine a torsion theory of $\underline{\mathcal{G}}$. Indeed it is the smallest localizing subcategory of $\underline{\mathcal{G}}$ containing all the simple objects.

By transfinite recursion we now define an ascending sequence of localizing subcategories of $\underline{\mathcal{G}}$:

$$\underline{\mathcal{G}}_0 \subset \underline{\mathcal{G}} \subset \cdots \subset \underline{\mathcal{G}}_\alpha \subset \cdots \subset \underline{\mathcal{G}}$$

such that $\underline{\mathcal{G}}_0 = \{0\}$, and $\underline{\mathcal{G}}_1$ is the localizing subcategory of all semi-Artinian objects of $\underline{\mathcal{G}}$ as defined above. If α is an ordinal such that for every $\beta < \alpha$ we have already defined $\underline{\mathcal{G}}_\beta$, then:

1. If α is not a limit ordinal, that is, we may view $\alpha = \beta + 1$, we write $\underline{\mathcal{G}}/\mathcal{G}_\beta$ for the quotient category of $\underline{\mathcal{G}}$ with respect to $\mathcal{G}_\beta$ and $\underline{\mathcal{Q}}_\beta : \underline{\mathcal{G}} \to \underline{\mathcal{G}}/\underline{\mathcal{G}}_\beta$ for the canonical functor, which is known to be an exact functor;

2. if α is a limit ordinal, then we let $\underline{\mathcal{G}}_\alpha$ be the smallest subcategory containing $\cup_{\beta<\alpha}\underline{\mathcal{G}}_\beta$; in other words $M < \underline{\mathcal{G}}_\alpha$ if and only if for any subobject $M + N$ of M such that $N \neq M$, M/N contains a nonzero subobject isomorphic to an object of $\cup_{\beta<\alpha}\underline{\mathcal{G}}_\beta$.

   The assumption that $\underline{\mathcal{G}}$ has a generator leads to the existence of an ordinal ξ such that $\underline{\mathcal{G}}_\xi = \underline{\mathcal{G}}_{\xi+1} = \ldots$ An object M of $\underline{\mathcal{G}}$ has **Gabriel dimension** if $M \in \underline{\mathcal{G}}_\alpha$ for some ordinal α; if α is the smallest ordinal for which $M \in \underline{\mathcal{G}}_\alpha$, then we say that M has Gabriel dimension α and we denote this by $\text{Gdim}(M) = \alpha$. In particular $\text{Gdim}(M) = 1$ means that M is semi-Artinian, and if $M = 0$, then we put $\text{Gdim}(M) = 0$.

3.4.2 Properties of Gabriel Dimension

1. If N is a subobject of M in $\underline{\mathcal{G}}$, then M has Gabriel dimension if and only if N and M/N have Gabriel dimension. In that case $\text{Gdim}(M) = \sup\{\text{Gdim}(N), \text{Gdim}(M/N)\}$.

2. Consider a family of objects in $\underline{\mathcal{G}}$, $(M_i)_{i\in\mathcal{A}}$, having Gabriel dimension; then $\coprod_{i\in\mathcal{A}} M_i$ has Gabriel dimension $\sup_{i\in\mathcal{A}}\{\text{Gdim}(M_i), i \in \mathcal{A}\}$.

3. An object M of $\underline{\mathcal{G}}$ having Krull dimension also has Gabriel dimension and $\text{Kdim}(M) \leq \text{Gdim}(M) \leq 1 + \text{Kdim}(M)$.

4. For a Noetherian object M of $\underline{\mathcal{G}}$ we have that $\text{Gdim}(M) = 1 + \text{Kdim}(M)$.

Proof
Cf. P. Gabriel [10], or reference [33]. □

Observe that statement 4 above tells us that Gabriel dimension is indeed not so different from Krull dimension, at least for a Noetherian object. In case $\underline{\mathcal{G}} = \underline{\mathcal{G}}_\xi$ and ξ is the smallest ordinal for which this happens, we say that $\underline{\mathcal{G}}$ has Gabriel dimension ξ.

Lemma 3.8
If $\underline{\mathcal{G}}$ has generator U, then $\underline{\mathcal{G}}$ has Gabriel dimension if and only if U has Gabriel dimension and in this case $\text{Gdim}(\underline{\mathcal{G}}) = \text{Gdim}(U)$.

Lemma 3.9
If M in $\underline{\mathcal{G}}$ is such that for every nonzero subobject X of M we have that M/X has Gabriel dimension, then M has Gabriel dimension and $\mathrm{Gdim}(M) \leq \alpha + 1$, *where* $\alpha = \sup\{\mathrm{Gdim}(M/X), X \text{ nonzero subobject of } M\}$.

Lemma 3.10
If M in $\underline{\mathcal{G}}$ has Krull, respectively Gabriel dimension, then for any ordinal $\alpha > 0$ there is a largest subobject $\kappa_\alpha(M)$ of M having Krull-, respectively Gabriel-, dimension at most α.

For detail and more about these dimensions we refer the reader to [11] or [33].

Exercise 3.3
Introduce Krull and Gabriel dimension for a skew topology (axioms A.1., ... ,A.9) and relate to the corresponding dimensions of the commutative shadow (e.g., for a virtual topology). The latter being a modular lattice, the generalization to such lattices as introduced in reference [32] can be used.

Exercise 3.4
If a category $\underline{\mathcal{C}}$ has a Grothendieck representation Rep, then for each object R in $\underline{\mathcal{C}}$ we have defined Gdim(Rep$R$)*, which we term the* ***representation dimension*** *of R. Develop a theory for this dimension in the category $\underline{\mathcal{C}}$ assuming suitable properties of the Grothendieck representation (e.g., measuring, ... , as in Sections 3.1 and 3.3). Relate the representing dimension to the one stemming from a Grothendieck quotient representation via the links between dimensions inherent in the notion of* ***topological nerve****.*

Exercise 3.5
Apply the notion of Gabriel dimension to sheaf categories and specific objects like structure sheaves (see techniques introduced in the next chapter too). Using Exercise 3.4, the notion of dimension may be applied to functor categories (and then they need not be Grothendieck categories but just a suitable kind of noncommutative lattice). Study behavior of the dimension with respect to separable functors between the underlying categories.

Exercise 3.6
The author is unaware of a possible application of Gdim to classes of modules over (higher-rank) valuation rings; here the absence of Noetherian conditions of the ring and most of its modules should provide extra interest in the use of Gabriel dimension even in the commutative case. Both the development of Gabriel dimension theory for modules over a commutative or over a noncommutative valuation ring may be worthwhile; the structure of not finitely generated modules over a nondiscrete valuation ring can be complex and almost pathological, but at least for valuation rings of

finite rank the Gabriel dimension could be expected to be controllable and therefore useful.

3.4.3 Project: General Birationality

Birationality between algebras, not necessarily commutative, has already been introduced and studied by the author. The definition depends on the existence of suitable localizations. Roughly speaking, one could say that finitely generated algebras A and B over a field K are birational if there exist kernel functors α, respectively β, on A-mod, respectively B-mod, such that $Q_{\kappa_\alpha}(A) \cong Q_{\kappa_\beta}(B)$ as K-algebras. Good ring theoretical behavior may be ensured by restricting to algebras A and B that are Goldie rings, such as prime Noetherian rings or Ore domains. On one hand, restrictions may be put on the kernel functor, such as being exact or being Ore localizations or even being central or normalizing localizations; on the other hand the condition of yielding isomorphic algebras may be weakened so that Morita equivalence is incorporated (recall that a kind of commutator condition has been introduced allowing isomorphism up to Azumaya algebra tensor factors in earlier work). There is certainly an interest in a well-behaved birationality in commutative geometry; in the noncommutative situation the aforementioned more recent results have dealt with birationality in terms of valuation rings or blowing-up and blowing-down. The general approach suggested in these notes allows a very general unifying theory, most of which is yet to be developed.

Starting from a category $\underline{C}$ with a suitable Grothendieck representation, objects A and B are said to be **birational** if there are kernel functors κ_A, respectively κ_B, on Rep(A), respectively Rep(B), such that the corresponding quotient categories are equivalent (could be strengthened to isomorphic). There is still a lot of freedom to adapt this definition; for example, if the Grothendieck representation is measuring, then one may ask that $Q_{\kappa_A}({}_AA) \cong Q_{\kappa_B}({}_BB)$, but the latter isomorphism is unnatural because a priori there need not even be a map between those localized objects. In case of spectral representations we may ask $A(\kappa_\alpha) \cong B(\kappa_\beta)$ in $\underline{C}$ (see Proposition 3.2), which is an object-wise demand weaker than categorical conditions like Rep($A(\kappa_\alpha)$) $\cong$ Rep($B(\kappa_\beta)$).

At first one tries to obtain a notion applicable in the noncommutative algebraic geometry, but it is worthwhile to go for a more abstract version using general noncommutative topology (and structure sheaves over them as in the next chapter); in particular, the idea above should then at least be extended to include nonidempotent radicals κ_A, κ_B (using torsion theories means working on the commutative shadow; of course even that provides new theory over noncommutative algebras and their birationality).

Chapter 4

Sheaves and Dynamical Topology

4.1 Introducing Structure Sheaves

Let us consider a category $\underline{\mathcal{R}}$ as in Section 3.1 together with a Grothendieck categorical representation rep : $\underline{\mathcal{R}} \to \underline{\mathcal{G}}$ as defined in Definition 3.1.

For an object R in $\underline{\mathcal{R}}$ we have Top(R) as defined before Definition 3.3, which may be viewed as a commutative topology (it is a modular lattice) of the object R. We also have the noncommutative topologies, $\Lambda(R)$, $T(R)$ respectively as Λ and T in Lemma 3.1. Recall from Corollary 3.3 that T is even a topology of virtual opens with a commutative shadow $SL(\Lambda(R))$ that is exactly Top(R).

Now consider a morphism $f : S \to R$ in the category $\underline{\mathcal{R}}$. To f we associated an exact functor, F : Rep(R) $\to$ Rep(S), which commutes with coproducts and defines a map F° : Top(S) $\to$ Top(R), also denoted by $F^\circ = \widetilde{f}$. This map F° respects the poset structure and is moreover continuous in the gen-topology. The importance of Proposition 3.6 is that the foregoing extends to the noncommutative structure constructed by considering hereditary preradicals; that is, we obtain a poset respecting map $Q_h(S) \to Q_h(R)$ with respect to the original operations $\wedge$ and $\vee$; in general we do **not** obtain a map $\Lambda(S) \to \Lambda(R)$ respecting the operation $\prod$.

Let $\underline{\mathcal{X}}$ be the class of noncommutative topologies (we restrict to the case of commutative $\vee$ here, such as virtual topologies) with morphisms that are poset morphisms respecting the operations $\vee$, $\wedge$ and continuous in the gen-topologies. A given Grothendieck representation $\underline{\mathcal{R}} \to \underline{\mathcal{G}}$ gives rise to several functors $\underline{\mathcal{R}} \to \underline{\mathcal{X}}$, as follows:

1. $\underline{\text{Top}}$ given by $\underline{\text{Top}}(R) = \text{Top}(R)$,
2. $\underline{T}$ given by $\underline{T}(R) = T(R)$,
3. $\underline{Q}_h$ given by $\underline{Q}_h(R) = Q_h(R)$.

In each of these cases it is clear how the functors act on morphisms. Observe that there are natural transforms $\underline{\text{Top}} \to \underline{T} \to \underline{Q}_h$.

So we are ready to define a category $\underline{\text{Top}}(\underline{\mathcal{R}})$ having for objects the Top(R) for R in $\underline{\mathcal{R}}$, and the morphisms Hom(Top(R), Top(S)) $= \text{Hom}_{\underline{\mathcal{R}}}(R, S)^\sim = \{\widetilde{f}, f \in \text{Hom}_{\underline{\mathcal{R}}}(R, S)\}$.

In a formally similar way we may define the categories $\underline{T}(\underline{\mathcal{R}})$ and $\underline{Q}_h(\underline{\mathcal{R}})$.

The GC representation Rep then yields the functors $\underline{\mathcal{R}} \to \underline{\text{Top}(\mathcal{R})}$, respectively $\underline{\mathcal{R}} \to \underline{T}(\underline{\mathcal{R}})$, resp. $\underline{\mathcal{R}} \to \underline{Q}_h(\underline{\mathcal{R}})$, defined by taking R in $\underline{\mathcal{R}}$ to Top(R) and $f : S \to R$ to $\widetilde{f}$: Top(S) $\to$ Top(R) for a morphism f in $\underline{\mathcal{R}}$, resp. taking R to $T(R)$ and f to $f^{\sim} : T(S) \to T(R)$, resp. taking R to $Q_h(R)$ and f to $\widehat{F} : Q_h(S) \to Q_h(R)$ (notation as in Lemma 3.2). Observe that $\widehat{F}$ restricts to $f^{\sim}$ on $T(S)$ and further to $\widetilde{f}$ on Top(S). We fix notation as follows: $X : \underline{\mathcal{R}} \to X(\underline{\mathcal{R}})$, meaning that X is one of the functors constructed above. We call X a **topologizing functor** and refer to $X(\underline{\mathcal{R}})$ as the **topoflow** along $\underline{\mathcal{R}}$.

Let us point out once more that when compared to classical notions of topology, Top(R) corresponds to Open^{op} as a poset. One should keep this in mind when presheaves and presheaf-flows are being defined in the following paragraph.

For R in $\underline{\mathcal{C}}$ we define an R-**presheaf** $P(R)$ with values in a category $\underline{\mathcal{D}}$ (sometimes referred to as being of type X, if $X(R)$ is a priori specified) as a functor $P(R)$: $X(R) \to \underline{\mathcal{D}}$.

Starting from a given Grothendieck representation Rep : $\underline{\mathcal{C}} \to \underline{\mathcal{G}}$ we define a **presheaf-flow** (of type X) along $\underline{\mathcal{C}}$ as a functorial morphism $P : X \to$ Rep, that is, a class of functors $P(X)$, $P(R) : X(R) \to$ Rep(R) for R in $\underline{\mathcal{C}}$, such that for every morphism $f : S \to R$ in $\underline{\mathcal{C}}$ we obtain a commutative functor diagram:

$$\begin{array}{ccc} X(S) & \xrightarrow{\widetilde{f}} & X(R) \\ \downarrow{\scriptstyle P(S)} & & \downarrow{\scriptstyle P(R)} \\ \text{Rep}(S) & \xleftarrow[F]{} & \text{Rep}(R) \end{array}$$

Let us write $X_{\text{gen}}(R)$ for the gen-topology defined on $X(R)$; recall that it has as a basis the sets $\{\lambda \in X(R), \tau \leq \lambda\}$ for $\tau \in X(R)\}$. In general, for any given basis $\mathcal{B}_R$ of $X_{\text{gen}}(R)$, a functor $X(R) \to \underline{\mathcal{D}}$, defines by restriction a functor $\mathcal{B}_R^0 \to \underline{\mathcal{D}}$ where $\mathcal{B}_R^\circ$ is viewed as a subcategory of $X(R)$ in the obvious way and B_R° appears because for $\tau \leq \tau'$ in $X(R)$ we get gen$(\tau') \subset$ gen(τ) in $X_{\text{gen}}(R)$. Under some extra gluing conditions the restricted functor $\mathcal{B}_R^\circ \to \underline{\mathcal{D}}$ would define a sheaf or at least a nice presheaf on $X_{\text{gen}}(R)$ (that is completely determined by its action on a basis of the topology). In general a given $P(R)$ **need not define** unambiguously a presehaf $P(R)_{\text{gen}} : X_{\text{gen}}(R)^\circ \to \underline{\mathcal{D}}$; we write $P(R)_{\mathcal{B}}$ for the functor $\mathcal{B}_R^\circ \to \underline{\mathcal{D}}$ defined by $P(R)$.

We may say that $P(R)$ is **gen-induced** if $P(R)_{\mathcal{B}}$ is in fact obtained as the restriction of a presheaf $P(R)_{\text{gen}} : X_{\text{gen}}(R)^\circ \to \underline{\mathcal{D}}$; however, when dealing with presheaves, a weaker condition involving only $\mathcal{B}_R$ is useful. Observe that a presheaf-flow P does induce a presheaf-flow $P_{\mathcal{B}} : \mathcal{B}^\circ \to$ Rep where $\mathcal{B}$ is the subfunctor of X obtained by selecting the canonical basis $\mathcal{B}_R$ in every $X(R)$ with morphisms behaving well because of the continuity of $\widetilde{f}$ in the gen-topologies of $X(S)$ and $X(R)$ for every $f : S \to R$ in $\underline{\mathcal{C}}$. With notation as in the diagram, we may define $P(S)_{*,\mathcal{B}}$ by taking for $U \in \mathcal{B}_R$:

$$P(S)_{*,\mathcal{B}}(U) := P(S)(\widetilde{f}^{-1}(U))$$

where for $U = \text{gen}(\tau)$, $\widetilde{f}^{-1}(U) = \text{gen}(\xi_\tau)$ and ξ_τ is obtained as in Corollary 3.1. For $T \to S \to R$ in $\underline{\mathcal{C}}$, compatibility of the diagrams as before is guaranteed. We now say that a presheaf-flow $P : X \to$ Rep is **genetic** if we have a natural transform of functors $P(S)_{*,\mathcal{B}} \to FP(R)_{\mathcal{B}}$ compatible with composition of morphisms in $\underline{\mathcal{C}}$.

4.1.1 Classical Example and Motivation

If $\underline{\mathcal{C}}$ is the category of commutative rings (or k-algebras over some field k), the classical construction of the spectrum and its Zariski topology correspond to the canonical topologization stemming from the Grothendieck representation $R \longrightarrow R$-mod, utilizing at several instances the fact that in this case the opens are determined by their points; that is, the prime ideals do form a quantum basis, the point topology is the gen-topology with respect to the dual poset structure, sheafification of presheaves is possible on the spectrum (= quantum base) and so forth. Moreover, a morphism between commutative rings not only induces a genetic (pre-)sheaf-flow of structure (pre-)sheaves but this happens also in a functorial way.

If $\underline{\mathcal{C}}$ is the category of not-necessarily commutative rings (or k-algebras over some field k) the behavior of Top, T or Q_h becomes different and we may still view Top as a kind of commutative topology because its properties correspond to those of a lattice. Consequently sheafification techniques exist subject to some modifications with respect to the commutative case but no real problems appear. The noncommutative topologies we introduced in Chapter 3 connected to the Grothendieck representations $R \to R$-mod, and in fact also the variations obtained in specific cases like the graded case $R \to R$-gr or the positively graded case $R \to \text{Proj}(R)$, are still satisfactory. However, ring morphisms $S \to R$ now still yield a (pre)sheaf-flow of structure (pre-) sheaves but functoriality is replaced by natural transform properties as continuity in the gen-topologies! In the presence of Rep-affine elements, for example, when a basis of such elements exists for the noncommutative topology considered, most aspects of the theory of abstract varieties or schemes survive in the noncommutative case. This provides the inspiration for defining abstract noncommutative spaces as in the next section.

4.1.2 Abstract Noncommutative Spaces and Schemes

We start from a given Grothendieck representation Rep : $\underline{\mathcal{R}} \to \underline{\mathcal{G}}$ as usual and consider an object Λ of $\underline{\mathcal{X}}$ as introduced before. We say that Λ is $\underline{(X, \text{Rep})}$**-covered** if there are finite sets $\{R_i, i\}$ in $\underline{\mathcal{R}}$ and $\{\lambda_i, i\}$ in Λ such that the latter is a global cover in Λ such that for every i, $[0, \lambda_i] \cong X(R_i)^\circ$; that is, these are poset isomorphic with respect to a bijection respecting $\wedge$ and $\vee$ (note that these then correspond to $\vee$ and $\wedge$ respectively in $X(R_i)$. We say that Λ is $\underline{(X, \text{Rep})}$**-pseudoaffine** if it is (X, Rep)-covered and for all $\mu \in \Lambda$ we have $\mu = \vee_i(\mu \wedge \lambda_i)$. Several strengthenings of this definition may be introduced, such as Λ is centrally (X, Rep)-pseudoaffine if the global cover $\{\lambda_i, i\}$ is central, that is, $\lambda \wedge \mu = \mu \wedge \lambda$ for all $\mu \in \Lambda$. Note that the second part of the definition is automatically fulfilled if axiom A.10 holds in Λ (we could restrict $\underline{\mathcal{X}}$ to such noncommutative topologies). If Λ is (X, Rep)-pseudoaffine, then it is said to be Noetherian-pseudoaffine if each $X(R_i)$ is Noetherian.

Proposition 4.1
If Λ is respectively Noetherian (X, Rep)-covered, Noetherian (X, rep)-pseudoaffine, Noetherian centrally (X, Rep)-pseudoaffine, then $C(\Lambda)$ is also.

Proof

Write A_i for the constant directed set defined by $\lambda_i \in \Lambda$ and look at $[A] \in C(\Lambda)$. We have:

$$[A] \wedge [A_i] = [A \wedge A_i] = [\{a \wedge \lambda_i, a \in A\}]$$

In case $a \wedge \lambda_i = \lambda_i \wedge a$ we obtain $[A] \wedge [A_i] = [A_i] \wedge [A]$. Now $\vee_i([A] \wedge [A_i])$ is the class of the filter on the set of elements $\vee_i(b_i \wedge \lambda_i)$ in Λ. Since the $b_i \in A$ vary over a finite index set we may choose $b' \in A$ such that $b' \leq b_i$ for all i and obtain $\vee_i(b' \wedge \lambda_i) \leq \vee_i(b_i \wedge \lambda_i)$. From the (X, rep)-cover condition for Λ we find that $b' = \vee_i(b' \wedge \lambda_i)$; hence $b' \leq \vee_i(b_i \wedge \lambda_i)$, or the latter is in the filter defining $[A]$. Thus $[A] \leq \vee_i([A] \wedge [A_i])$ and therefore equality follows. In particular it is clear that $\{[A_i], i\}$ is a global cover for $C(\Lambda)$. If $[B] \leq [A_i]$, then there is a $b \in B$ such that $b \leq \lambda_i$. Let us denote the bijection $[0, \lambda_i] \longrightarrow X(R_i)^\circ$ by γ_i; then to $b \in B$ as before there corresponds $\gamma_i(b)$. To a downward directed set B we may associate $\vee_{b \in B} \gamma_b$ in $X(R_i)$. Observe that the family $\mathcal{F}_B^i = \{\gamma_i(b), b \in B, b \leq \lambda_i\}$ has a unique maximal element; indeed maximal elements exist in view of the Noetherian assumption, and if $\gamma_i(b_1)$ and $\gamma_i(b_2)$ are two such elements, then for some $b_3 \in B$ such that $b_3 \leq b_1$ and $b_3 \leq b_2$ we have $\gamma_i(b_1) \leq \gamma_i(b_3)$ and $\gamma_i(b_2) \leq \gamma_i(b_3)$; hence the maximality assumption entails $\gamma_i(b_1) = \gamma_i(b_3) = \gamma_i(b_2)$. Let us write $\gamma_i(b_0)$ for the maximal element of $\mathcal{F}_B^i$. If $[B_1] \underset{\neq}{<} [B_2] \leq [A_i]$; then for every $b_2 \in B_2$ there exists a $b_1 \in B_1$ such that $b_1 \leq b_2$ and the strictness yields that some $b_1' \in B$ is not in the filter of B_2. If the maximal elements of respectively $\mathcal{F}_{B_1}^i$, $\mathcal{F}_{B_2}^i$ are the same, then there is a $b_1' \in B_1$, $b_1' \leq \lambda_i$ not in the filter of B_2; hence $\gamma_i(b_1') \leq \gamma(b_{10}) = \gamma_i(b_{20})$ (the latter denoting the maximal elements of $\mathcal{F}_{B_1}^i$, resp. $\mathcal{F}_{B_2}^i$), where $b_{10} \in B_1$ and $b_{10} \leq \lambda_i$, $b_{20} \in B_2$, $b_{20} \leq \lambda_i$. Bijectivity of γ_i then yields $b_{20} \leq b_1'$, contradicting the fact that b' is not in the filter associated to B_2. Hence γ_i induces a bijection $[0, [\lambda_i]] \to X(R_i)^\circ$.

□

Again starting from a Grothendieck representation Rep as usual, restricting to one having suitable properties in practical situations, for example, assuming it to be spectral or induced schematic, we may define abstract affine Λ as follows.

Definition 4.1

A $\Lambda = \Lambda(R)$ for $R \in \underline{R}$, that is, (X, Rep)-pseudoaffine as before, is said to be **affine** if for $\gamma \in \Lambda$ and any $M \in \text{Rep}(R)$, the canonical map in $\text{Rep}(R)$

$$\rho_\gamma^1(M) : M \longrightarrow Q_\gamma(M)$$

is the pullback of the system of maps (in $\text{Rep}(R)$)

$$\rho_{\gamma \wedge \lambda_i}^1(M) : M \longrightarrow Q_{(\gamma \wedge \lambda_i)}(M)$$

where the λ_i vary over the global cover $\{\lambda_i, i\}$ in Λ used in the definition of the pseudoaffine property, and each λ_i is Rep-affine as in Definition 3.6. The latter justifies the use of Q_γ as explained in the remark preceding Proposition 3.11.

Proposition 4.2
If Rep is inductive schematic and Λ is pseudoaffine with respect to a cover $\{\lambda_i, i \in \mathcal{I}\}$ such that each λ_i is Rep-affine, then Λ is affine.

Proof
In view of Proposition 3.11, Q_{λ_i} is exact and commutes with coproducts, thus any $\kappa \leq \lambda_i$ corresponds to a torsion theory that is compatible with Q_κ (see definition before Lemma 2.23 as well as Proposition 2.27 and Corollary 2.12). In particular this holds for $\gamma \wedge \lambda_i; i \in J$ and moreover the pseudoaffinity of Λ entails that $\gamma = \vee_i(\gamma \wedge \lambda_i)$ and we may identify $[0, \lambda_i]$ to $X(R_i)^{\mathrm{op}}$, associating γ_i to $\gamma \wedge \lambda_i$. Since Rep is induced schematic $R_i \cong Q_{\lambda_i}(R)$ and $Q_{\gamma_i}(R_i) \simeq Q_{\gamma_i} Q_{\lambda_i}(R) \cong Q_{(\gamma\wedge\lambda_i)}(R)$ where $\gamma \wedge \lambda_i$ in Λ corresponds to the radical $\gamma_i \in X(R_i)$ (we write $(\gamma \wedge \lambda_i)$ as index to Q in order to indicate that there is an opposite in the relation between $X(R_i)$ and $[0, \lambda_i]$!) The exactness of Q_{λ_i} entails $Q_{\gamma_i} Q_{\lambda_i} = Q_{\lambda_i} Q_{\gamma_i} = Q_{\lambda_i} Q_{(\gamma)}$. Since $\{\lambda_i, i\}$ is a global cover, the map $\rho^1_\gamma(M)$ is indeed the pullback for the system $M \to Q_{(\gamma\wedge\lambda_i)}(M) = Q_{\lambda_i} Q_{(\gamma)}(M)$ (where (γ) refers to the radical correspondence to γ as an element of Λ). □

Proposition 4.3
If Λ is Noetherian, and affine, then $C(\Lambda)$ is affine.

Proof
As in Proposition 4.2 we derive that $C(\Lambda)$ is pseudoaffine; the statement follows from the compatibility of direct limits and the pullbacks (as inverse limits over the cover considered). □

Examples of "abstract" noncommutative affine Noetherian Λ can be found by considering the gen-topology on the torsion theoretic prime spectrum of a Noetherian schematic algebra, for example, the Weyl algebra, quantum Sl_2, generalized gauge algebras and the Witten algebra, and so forth. The projective or graded version of the torsion theoretic prime spectrum, now of a suitably graded ring (which is different from the categorically defined Proj, a priori), or other generalizations of prime spectra of similar type, are again examples, where the GC representations are obtained from (graded) modules. In fact, it is possible to construct abstract spaces by constructing a bunch of functors suitably stacked together; special examples may be obtained by putting gluing conditions such that the *scheme* of functors is covered by opens that look like an *affine scheme* of localization functors of some nice category. As an example that need not make reference to localization theory directly, let us mention the case of **essential functors** in some detail (see also [46]).

We consider a category $\underline{\mathcal{C}}$ allowing products and co-products, limits and co-limits, so that we may perform suitable pullbacks and pushout operations. Even though this is not really necessary, we may think of categories $\underline{\mathcal{C}}$ having additive groups underlying its objects. Let $E(\underline{\mathcal{C}})$ be the class of left exact endofunctors $F : \underline{\mathcal{C}} \to \underline{\mathcal{C}}$, and we may view natural transforms of functors as morphisms in $E(\underline{\mathcal{C}})$ allowing us to consider $E(\underline{\mathcal{C}})$ as a category again.

An F of $E(\underline{\mathcal{C}})$ is an **essential functor** whenever there is a natural transform j from the identity functor I to F, denoted by $j : I \to F$, such that for every morphism $M \to N$ in the category $\underline{\mathcal{C}}$, the morphism $FM \to FN$ in $\underline{\mathcal{C}}$ is the **unique** one making the following diagram commutative:

$$\begin{array}{ccc} M & \xrightarrow{f} & N \\ {\scriptstyle j_M}\downarrow & & \downarrow{\scriptstyle j_N} \\ FM & \xrightarrow{F_f} & FN \end{array}$$

The essential functors define a subclass $EE(\underline{\mathcal{C}})$, which is closed under composition. Indeed, the following commutative diagram

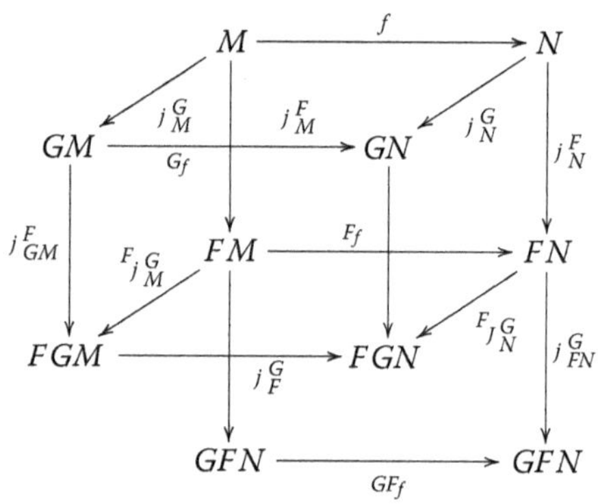

where $j^F : I \to F$, $j^G : I \to G$ are natural transforms, we have uniqueness of Gf and Ff by essentiality of G, respectively F, and uniqueness of FGf and GFf by essentiality of F, resp. G.

Of course the composition

$$M \xrightarrow{j^F_M} FM \xrightarrow{j^G_{FM}} GFM$$

defines the natural transform $I \to GF$; this is similar for $I \to FG$. We define $F \leq G$ **for essential functors** F and G by the existence of a natural transform t: $F \to G$ such that we have a functorial commutative triangle:

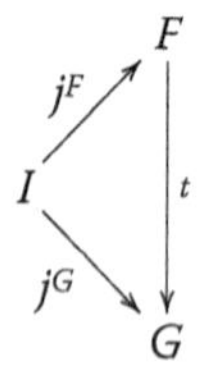

That is, for $f : N \to M$ in $\underline{C}$ we have a commutative diagram

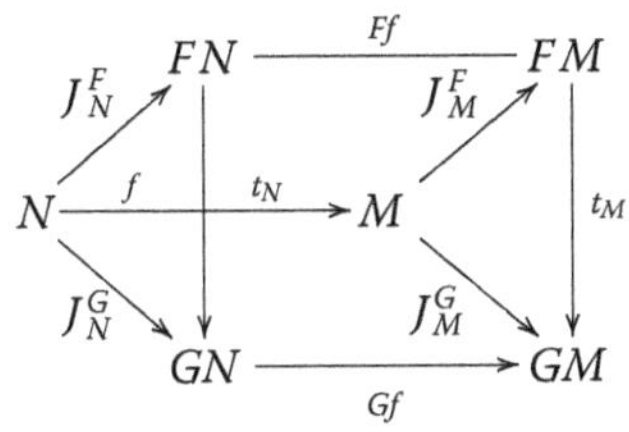

Restricting attention to the essential functors in $E(\underline{C})$ we thus arrive at a poset $EE(\underline{C})$, which can also be seen as a subcategory of $E(\underline{C})$ with respect to the stucture defined above. For given F and G in $EE(\underline{C}|)$ we obtain new functors FG and GF in $EE(\underline{C})$ and for every M in $\underline{C}$ we obtain a diagram:

$D(M)$

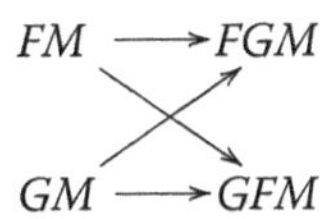

The assumptions on the category $\underline{C}$ yield the existence of a pullback for the foregoing diagram; we shall denote it by $(F \wedge G)M$; observe that it does not depend on the ordering of F and G! If $f : N \to M$ in $\underline{C}$, then we obtain morphisms in $\underline{C}$ fitting in the diagram $D(f) : D(N) \to D(M)$ so that classical properties of the pullback construction yield the existence of a morphism $(F \wedge G)f : (F \wedge G)N \to (F \wedge G)M$. It is easy to see that $F \wedge G$ is a left exact endofunctor of $\underline{C}$ and moreover, essentiality of F and G yields a natural transform $I \to F \wedge G$ making the following diagram of functorial morphisms commutative

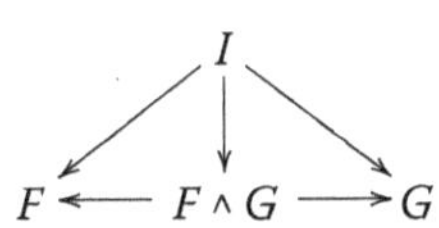

where $F \wedge G \to G$, respectively $F \wedge G \to F$, derives from $(F \wedge G)M \to GM$ appearing in the pullback diagram for $D(M)$, respectively $(F \wedge G)M \to FM$. Obviously $F \wedge G \leq F$ and $F \wedge G \leq G$, but it is not necessarily so that $F \wedge G$ is also automatically an essential functor. The latter does follow, for example, when the canonical morphisms $(F \wedge G)N \to FN$, or the $(F \wedge G)N \to GN$, are monomorphisms. In any case, for a noncommutative topology Λ we may define a **Λ-spectrum** for $\underline{C}$ as a poset morphism $\Lambda^\circ \to EE(\underline{C})$ that does respect $\wedge$ and $\vee$; that is, a Λ-spectrum for $\underline{C}$ may be viewed as a presheaf over Λ with values in the functor category $EE(\underline{C})$ respecting the noncommutative operations. If Γ is a Λ-spectrum on $\underline{C}$, then an idempotent λ in Λ corresponds to an idempotent functor $\Gamma(\lambda)$. Without going into the details, let us just claim here that a Λ-spectrum may be extended to a $C(\Lambda)$-spectrum by using *direct limits* in $EE(\underline{C})$, and further to the pattern topology generated by the idempotents. As a general definition of a structure sheaf we may now establish the following definition.

Definition 4.2
Let Γ be a Λ-spectrum on $\underline{\mathcal{C}}$; then a Γ**-structure (pre)sheaf** over Λ with values in $\underline{\mathcal{C}}$ is a (pre)sheaf P over Λ with values in $\underline{\mathcal{C}}$ such that it factors over Γ in the sense that for all $\lambda \in \Lambda$ we have $P_\lambda = \Gamma(\lambda)M$ for some object M of $\underline{\mathcal{C}}$.

Theorem 4.1
With notation and convention as above: Γ-structure presheaves are sheaves.

Proof
A cover $\lambda = \lambda_1 \vee \ldots \vee \lambda_n$ in Λ yields the functor relation: $\Gamma(\lambda) = \Gamma(\lambda_1) \wedge \cdots \wedge \Gamma(\lambda_n)$; note that the functor on the righthand side is now forced to be essential because $\Gamma(\lambda)$ is by assumption. For every (i, j) with $i, j \in \{1, \ldots, n\}$ we obtain pullback diagrams in $\underline{\mathcal{C}}$, for some $M \in \underline{\mathcal{C}}$:

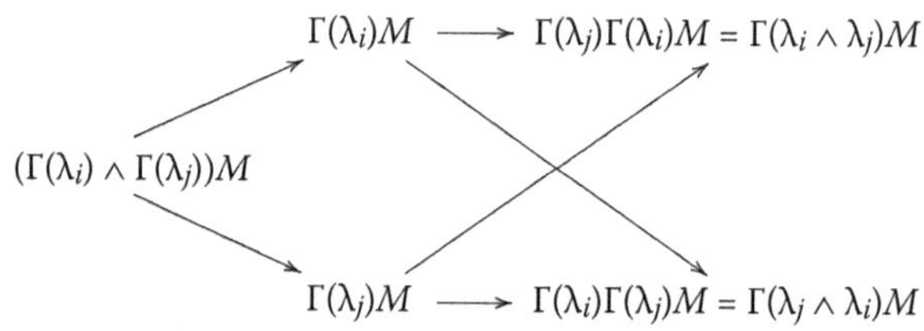

Consequently, the sheaf properties are now embedded intrinsically in the construction because properties were immediately phrased on the functorial level. □

Corollary 4.1
The structure presheaves defined earlier provide examples of an application of the foregoing theorem.

i. *Structure (pre)sheaves of modules over the prime spectrum of a commutative ring.*

ii. *Structure (pre)sheaves of modules over a noncommutative ring either over the prime spectrum (cf. [30], [44]), or over the torsion theoretic spectrum (cf. [14], [48]). In particular over rings satisfying polynomial identities (cf. [47], [37]).*

iii. *Structure (pre)sheaves over the noncommutative topology and micro-structure (pre)sheaves in particular over rings of differential operators (cf. [38], [40], [27]).*

iv. *Structure (pre)sheaves over noncommutative projective spaces (cf. [4], [14], ...) and structure (pre)sheaves of so-called quantum sections (cf. [24], [40]).*

4.1.3 Project: Replacing Essential by Separable Functors

Instead of essential functors we may consider separable functors and follow this through in the foregoing section. One may prefer to work on a noncommutative Grothendieck topology and consider a suitable bunch of separable functors over it. How far can one develop the noncommutative version of étale covers (separable covers seen as a generalization of Zariski covers which correspond to suitable bunches of localization or essential functors) and later étale cohomology? Look also at the cohomology project (see 4.2.3) in the dynamical theory.

4.1.4 Example: Ore Sets in Schematic Algebras

Let us return to the very first example of a noncommutative topology and its sheaf theory to be considered. We let $\underline{\mathcal{W}}$ be the noncommutative Grothendieck topology as considered in Section 2.8.; $\underline{\mathcal{W}}$ is associated to a connected positively graded K-algebra A, $A = K \oplus A_1 \oplus \ldots \oplus A_n \oplus \ldots$, and we assume that $A = K < A_1 >$. Everything we state in this section may be easily modified so that it holds for the virtual topology defined by $\underline{W}$ (i.e., generated by Ore localization functors).

We look at a presheaf $\mathcal{Q} : \underline{\mathcal{W}} \to A\text{-gr}$; that is, for all $W \in \underline{\mathcal{W}}$ the sections $\mathcal{Q}(W)$ of $\mathcal{Q}$ over W form a graded $Q_S(A)$-module where S stands for the last letter of W. If A is commutative, this definition reduces to Q being a presheaf of graded $\underline{O}^g_Y$-modules where $\underline{O}^g_Y$ is the graded structure sheaf over $Y = \text{proj}(A)$; by passing to homogeneous parts of degree zero we arrive at the usual structure sheaf $\underline{O}_y$ of Y. For $W = 1$ we demand $\mathcal{Q}(1)$ to be a $Q_{\kappa_+}(A)$-module and we write $\Gamma_*(\mathcal{Q} = \mathcal{Q}(1)$. We write

$$\rho^W_V : \mathcal{Q}(W) \to \mathcal{Q}(V)$$

for the restriction morphism with respect to $V \to W$ in $\underline{\mathcal{W}}$. If $W = 1$, then we simply write $\rho_V : \Gamma_*(\mathcal{Q}) \to \mathcal{Q}(V)$.

The sheaf axioms are now phrased as follows:

i. For $W \in \underline{\mathcal{W}}$ and any global cover $\{W_i, i \in J\}$, $\rho^W_{W_i W}(m) = 0$ in $\mathcal{Q}(W_i W)$ for all $i \in J$, yields $m = 0$.

ii. For $W \in \underline{\mathcal{W}}$ and global cover $\{W_i, i \in J\}$, if $m_i \in \mathcal{Q}(W_i - W)$ for $i \in J$ are such that for all $i, j \in J$:

$$\rho^{W_i W}_{W_i W_j}(m_i) = \rho^{W_j W}_{W_i W_j W}(m_j),$$

then there is $m \in \mathcal{Q}(W)$ such that $\rho^W_{W_i W}(m) = m_i, i \in J$. This m is unique in view of i.

A global cover $\{W_i, i \in \mathcal{J}\}$ and a $W \in \underline{\mathcal{W}}$ defines a full subcategory of $\underline{\mathcal{W}}$ consisting of $W_i W$, $W_i W_j W$, and we obtain an inverse system:

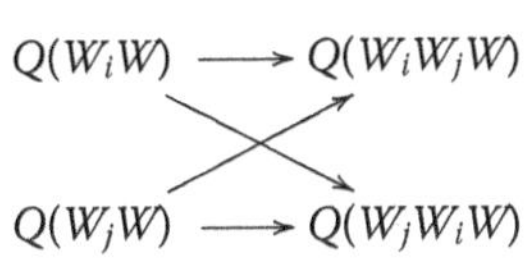

The graded inverse limit of this system is denoted by $\varprojlim_{i,j} \mathcal{Q}(W_i W)$ (a single index after the double indexed lim is not confusing because an element of the inverse limit is determined by its components in the $\mathcal{Q}(W_i W)$).

The presheaf $\mathcal{Q}$ is a sheaf exactly when for every global cover the inverse limit of the above system is isomorphic to $\mathcal{Q}(W)$ for every $W \in \underline{\mathcal{W}}$.

To any graded A-module M we associate a structure presheaf $\underline{O}^g M$ defined as follows:

$$\underline{O}^g_M(1) = Q_{\kappa_+}(M), \underline{O}^g_M(W) = Q_W(M) \text{ for } W \neq 1 \text{ in } \underline{\mathcal{W}}(A)$$

and to $V \to W$ there corresponds $\rho^W_V : Q_W(M) \to Q_V(M)$, which is the canonical graded morphism fitting in the commutative diagram:

$$\begin{array}{ccc} Q_W(M) & \xrightarrow{\rho^W_V} & Q_V(M) \\ & {\scriptstyle j_W}\nwarrow \quad \nearrow{\scriptstyle j_V} & \\ & M & \end{array}$$

Remember that for $M =_A A$ the sections over arbitrary W are not necessarily rings (this happens exactly when $\mathcal{L}(W)$ is idempotent, see Section 2.8). The equivalent of Theorem 4.1 is as follows.

Theorem 4.1: Rephrased

Let A be a schematic K-algebra as before; then $\underline{O}^g_M$ and $\underline{O}_M = (\underline{O}^g_M)_0$ are sheaves!

Of course, the original proof for this is independent of Theorem 4.1; see for example [47] where the noncommutative version of J. P. Serre's global section theorem was established for the first time.

We have so far obtained a noncommutative scheme structure on $Y = \text{proj}(A)$, denoted $\underline{O}_Y$, as follows: to $W = 1$ we associate the open $Y(1) = (A, \kappa_+)\text{-}gr_f$, that is, the whole proj(A)*; to $W \neq 1$ we associate the open $Y(W)$ defined as the class of graded A-modules of the form $Q_W(A) \otimes_A M$, $M \in A\text{-}gr_f$; in particular to an Ore set in $O(A)$ we associate a basic open $Y(S)$ given as $Q_S(A)\text{-}gr_f$ and this is equivalent to $Q_S(A)_0\text{-}mod_f = A_{(S)}\text{-}mod_{\in f}$ in view of the remark on affine sets in Section 2.8. On a basic open set $Y(S)$ we do have that $Q_S(A)$ is a (strongly) graded ring and $Q_S(M)$ is a graded $Q_S(A)$-module, but these statements do not hold over a general $Y(W)$. So we may sometimes want to restrict to affine opens and also affine covers, that is, a cover given by a finite number of Ore sets $\{T_i, i \in \mathcal{I}\}$ such that $\cap_i \mathcal{L}(T_i) = \mathcal{L}(\kappa_+)$, or $\kappa_+ = \kappa_{t_1} \wedge \ldots \wedge \kappa_{T_n}$.*

We say that a sheaf $\mathcal{S}$ on $\underline{\mathcal{W}}$ is **quasi-coherent** *if there exists an affine cover $\{T_i, i \in J\}$ for $Y =$* proj(A) *together with grade $Q_{T_i}(A)$-modules M_i such that for $V \to W$ in $\underline{\mathcal{W}}$ we obtain a commutative diagram in A-gr, the vertical maps being* *isomorphisms:*

$$\begin{array}{ccc} S(T_iW) & \longrightarrow & S(T_iV) \\ \downarrow & & \downarrow \\ Q_W(M_i) & \longrightarrow & Q_V(M_i) \end{array}$$

A quasi-coherent $\mathcal{S}$ is coherent if all M_i are finitely generated $Q_{T_i}(A)$-modules. Observe that only demanding isomorphisms $\mathcal{S}(T_i W) = Q_W(M_i)$ and not $Q_W(M_i) = \mathcal{S}(W)$ for every word W containing the letter T_i is essential!

If M is a (finitely generated) graded A-module (we keep assumptions on A as before), then $\underline{O}^g_M$ is (coherent) quasi-coherent.

Theorem (noncommutative version of Serre's Global Section Theorem): If $\mathcal{S}$ is a quasi-coherent sheaf on $\underline{\mathcal{W}}$ and $\Gamma_*(\mathcal{S})$ denotes the global section A-module, then $\mathcal{S}$ is isomorphic to the structure sheaf of $\Gamma_*(\mathcal{S})$. The category of quasi-coherent sheaves on $\underline{\mathcal{W}}$ is equivalent to the quotient category (A, κ_+)-gr of A-gr; the category of coherent sheaves on $\underline{\mathcal{W}}$ is equivalent to Proj(A) (i.e., $((A, \kappa_+)$-gr$_f$ by restricting to finitely generated objects). For the proof and more detail we refer the reader to [44] and [47]. For explicit calculation of sections and quantum sections of concrete algebras the reader should consult Section 2.3 in [44].

4.2 Dynamical Presheaves and Temporal Points

In this section we present some remarks of a rather esoteric nature but related to possible practical implementation of noncommutative geometry in physics. Indeed, it seems that certain branches of physics would benefit from a construction of a space continuum without time, or with time only present in a rudimentary form, for example, a step in the ordering of growing posets (quantum gravity from a poset point of view). Even more controversial ideas view space as materially created by events happening, so that space may be thought of as flowing in emptiness, making things happen; still another theory aims to construct space from information exchanges to arrive at "It from Bit." Well, in physics a model is only good until it becomes bad sooner or later.

It is my (poor) understanding that fundamental issues in the description of the material world arise from an incompatibility between a model that is intrinsically discrete because it is based on the finiteness of physical events and the mathematical description in terms of spaces and laws defined over real numbers. The possibility of measuring by numbers may be thought of as an unphrased axiom, but it may also be just a dream: perhaps real numbers are far too many, they may vary too continuously, and they may even not exist within the so-called physical reality one seeks to describe. The problem of "existence" of geometrical objects also appears in noncommutative geometry in the shape of the possible "lack of points."

The shortage of points is the reason that a noncommutative topology cannot be fit in set theory; therefore the shortage of points, that is, not every open is defined by its points, is equivalent to the noncommutativity. Nevertheless, noncommutative geometry could be an ingredient in understanding certain problems of synergy between the discrete and the continuous, for example, the noncommutative spaces modulo their commutative shadow are discrete, in other words, space combines noncommutative discreteness over commutative continuity in this description.

At this point one could follow either of two lines of development that may bring new ideas in the understanding of the relation between so-called physical reality and mathematical description.

1. Construct "space" as a **dynamical noncommutative topological** space and define geometrical objects as existing over some parameter (time) intervals. That is, there may not exist enough points on a given moment (fixing a parameter value), but there do exist enough points over a suitable parameter interval. This is philosophically satisfying; noncommutative continuity is introduced via the variation of an external parameter (why don't we agree to call it "time" from now on) but momentary observations, which are only abstractly possible (real measurements take time!), put us in the discrete-versus-continuous situation of noncommutative geometry.

   At the level of foundations, dynamical mathematical theories may have been neglected; nevertheless, I believe such theories fit the description of natural phenomena better, and application of such well-founded dynamical theories might make the development of perhaps less well-founded physical models superfluous!

2. Replace points, more precisely functions, defined in the set-theoretic spirit, by a generalization of "germs of functions" obtained by extending limit constructions in classical topology terms to noncommutative-type structures. This leads to a notion of a point as an atavar of "stalk" of a pregiven sheaf and becomes variable when different sheaves over the base noncommutative topology are considered. Assuming that the right (noncommutative) topological space and the correct sheaf of functions on it have been identified in order to describe some natural or geometric phenomenon, then the notion of points via stalks should be suitable too. For example, prime ideals would be identified via stalks if the structure sheaf of a commutative Noetherian ring would be pregiven, without having to check a primeness condition of the corresponding localization.

4.2.1 Project: Monads in Bicategories

Both Grothendieck's locales and the noncommutative sites developed here may be understood as monads in a bicategory. Construct the n-category version of noncommutative spaces abstractly as a pair of monads, one being the "commutative shadow" of the other. This commutative shadow is now a locale determined by commutation properties of morphisms. For more detail and basic theory we refer to [15]. The sheaf theory as well as the generalized germs of functions may be phrased in this context, so we do not go deeper into the possibilities relating to possibility 2 as above here.

In Section 2.1.3 we gave a definition for poset dynamics starting from a totally ordered poset T together with poset maps $\varphi_{tt'} : \Lambda_t \to \Lambda_{t'}$ for every $t \leq t'$ in T where now we assume the Λ_t to be a family of noncommutative topological spaces. The axioms DP1 $\to$ DP5 introduce aspects of continuity on the *index level*. We now introduce a notion of dynamical topology DT independent of the axioms DP. The definition of noncommutative space continuum is then obtained by combining the DT and DP axioms. For now we fix notation as above. If $A_i \subset \Lambda_t$ is a directed set, then for every $t', t \leq t'$, the set $A_{t'} = \varphi_{tt'}(A_t)$ is directed too; indeed, for given $\varphi_{tt'}(a), \varphi_{tt'}(b)$ in $\varphi_{tt'}(A_t)$ there exists a $c \leq a, b$, hence $\varphi_{tt'}(c) \leq \varphi_{tt'}(a), \varphi_{tt'}(b)$. In case directed sets A and B in Λ_t are equivalent we have, for every $t \leq t'$, that $\varphi_{tt'}(A)$ and $\varphi_{tt'}(B)$ are equivalent. Indeed, if $\varphi_{tt'}(a) \in \varphi_{tt'}(A)$, then there are b and b' in B such that $b \leq a \leq b'$, hence $\varphi_{tt'}(b) \leq \varphi_{tt'}(a) \leq \varphi_{tt'}(b')$ and similarly when the role of A and B is interchanged. Consequently, every $\varphi_{tt'}$ determines uniquely $\varphi^e_{tt'} : C(\Lambda_t) \to C(\Lambda_{t'})$ by $\varphi^e_{tt'}([A]) = [\varphi_{tt'}(A)]$. If $\varphi_{tt'}$ respects the operations $\wedge$ and $\vee$ in Λ_t, then $\varphi^e_{tt'}$ respects the operations $\wedge$ and $\vee$ of $C(\Lambda_t)$; indeed, for example:

$$\begin{aligned}
\varphi^e_{tt'}([A] \cap [B]) &= \varphi^e_{tt'}([A \dot{\wedge} B]) \\
&= [\varphi_{tt'}(A \dot{\wedge} B)] \\
&= [\varphi_{tt'}(A) \dot{\wedge} \varphi_{tt'}(B)] \\
&= [\varphi_{tt'}(A)] \wedge [\varphi_{tt'}(B)]
\end{aligned}$$

and similar with respect to $\vee$.

Lemma 4.1

The system of poset maps $\varphi_{tt'}$, $t \leq t'$ in T defines a system $\varphi^e_{tt'}$; if the maps $\varphi_{tt'}$ respect the operations $\wedge$ and $\vee$ of the noncommutative topologies Λ_t, then $\varphi^e_{tt'}$ does the same for $C(\Lambda_t)$. Moreover, in the latter situation $\varphi_{tt'}$ maps $\wedge$-idempotents of Λ_t to $\wedge$-idempotents of $\Lambda_{t'}$, respectively $\vee$-idempotents of Λ_t to $\vee$-idempotents of $\Lambda_{t'}$; if $[A_t]$ is strongly idempotent in $C(\Lambda_t)$, then $[\varphi_{tt'}(A_t)]$ is strongly idempotent in $C(\Lambda_{t'}]$.

Proof

The first statements are obvious from the foregoing remarks. If $\lambda \in \Lambda_t$ is idempotent, then we have:

$$\varphi_{tt'}(\lambda) \wedge \varphi_{tt'}(\lambda) = \varphi_{tt'}(\lambda \wedge \lambda) = \varphi_{tt'}(\lambda).$$

We have to check that $\varphi_{tt'}(A_t)$ is idempotently directed if A is such, so look at some $\varphi_{tt'}(a)$ for $a \in A_t$. Then there exists a $\lambda \in \mathrm{id}_\wedge(A_t)$, $\lambda \leq a$, hence $\varphi_{tt'}(\lambda) \in \mathrm{id}_\wedge(\varphi_{tt'}(A_t))$ and $\varphi_{tt'}(\lambda) \leq \varphi_{tt'}(a)$. $\square$

An element $\Lambda_t \in \Lambda_t$ not comparable to any element of Λ_t different from 0 and 1 is said to be **isolated**; the others are **interacting**. If we look at the graph of $\leq$ on $\Lambda_t - \{0, 1\} = \Lambda_t^*$, then isolated elements correspond to singleton connected components of the graph. The following axioms for DNT will imply that $\cup_{t'}\{\mathrm{Im}\varphi_{t't}, t' \leq t\}$ is exactly the set of interacting elements of Λ_t.

Corollary 4.2

For each Λ_t, let τ_t be the corresponding pattern topology and let $\{\varphi_{tt'}, t \leq t'$ in $T\}$ be a system of poset maps respecting the operations $\wedge$ and $\vee$. Write $\psi_{tt'}$ for the restriction *of $\varphi^e_{tt'}$ to τ_t for every $t \le t'$ in T; then the system of poset maps $\psi_{tt'} : \tau_t \to \tau_{t'}$ again preserves operations $\wedge$ and $\vee$.*

Proof

The pattern topology τ_t is obtained by taking all $\wedge$-finite bracketed espressions with respect to $\wedge$ and $\vee$ in the letters of $\mathrm{Id}_\wedge(C(\Lambda_t))$. The lemma entails that $\varphi^e_{tt'}$ maps τ_t to $\tau_{t'}$ and the claim follows. □

In the generality as above the maps $\varphi_{tt'}$ do not map points of Λ_t to points of $\Lambda_{t'}$; neither does $\varphi_{tt'}$ respect the operation $\underset{\cdot}{\wedge}$ of the commutative shadow $SL(\Lambda_t)$.

Definition 4.3: Dynamical Noncommutative Topology

As before, we consider a family of noncommutative topologies indexed by a totally ordered set T, together with poset-maps $\varphi_{tt'}$ for every $t \le t'$ in T, $\varphi_{tt'} : \Lambda_t \to \Lambda_{t'}$. This system is called a *dynamical noncommutative topology* (DNT) if the following conditions hold.

DNT1 For all $t \in T$, $\varphi_{tt} = I_{\Lambda_t}$. Also we assume that $\varphi_{tt'}(0) = 0$, $\varphi_{tt'}(1) = 1$.

DNT2 For $t \le t' \le t''$ in T, $\varphi_{t't''}\varphi_{tt'} = \varphi_{tt''}$ and $\varphi_{tt'}, t \le t'$ in T, preserves the operations $\wedge$ and $\vee$.

DNT3 (See Remarks after DP5, in 2.3.1.) If we have $x < y$ in Λ_t for some $t \in T$, then there is a $t < t_1$ in T such that for $z_1 \in \Lambda_{t_1}$ satisfying a nontrivial $\varphi_{tt_1}(x) < z_1 < \varphi_{tt_1}(y)$ there exists a $z \in \Lambda_t, x < z < y$, for which $\varphi_{tt_1}(z) = z_1$. Of course this is the stronger version of DP3 (as after DP5) but for $\mathcal{F} = \{t_1\}$, and we have then weakened it by only stating the existence of such $\mathcal{F}$ hence allowing a dependence on x, y (in other words, the existence of a strict intermediate element has a past but only within a certain "time" variation).

DNT4 (DNP4 and DNP5) For $t \in T$ and nontrivial $x < z < y$ (this means that we exclude $x = 0$ and $y = 1$) in Λ_t there exist t_1, t_2 in T such that $t_1 < t < t_2$ and for every $t' \in]t_1, t_2[$ we have either $t \le t'$ and $\varphi_{tt'}(x) < \varphi_{tt'}(z) < \varphi_{tt'}(y)$ or $t' \le t$, and if $x' < y' \in \Lambda_{t'}$ exist such that $\varphi_{t't}(x') = x$, $\varphi_{t't}(y') = y$, then there also exists z' in $\Lambda_{t'}$ such that $x' < z' < y'$ and $\varphi_{t't}(z') = z$.

Consequently, a nontrivial relation $x < y$ in Λ_t also lives in a T-interval containing t. The effect of the modification when compared to DP3 is that now for any interval $[x, y]$ in Λ_t, there is a $t_1, t < t_1$ such that $\varphi_{tt_1}([x, y])$ is the whole interval $[\varphi_{tt_1}(x), \varphi_{tt_1}(y)]$ in Λ_{t_1}, yet $\mathrm{Im}\varphi_{tt_1}$ is not convex in Λ_{t_1} because we allow t_1 to depend on $[x, y]$.

DNT5 In the foregoing we have seen that a relation $x < y$ in $\Lambda_t, t \in T$, stays alive in an open interval in T containing t; it is natural to demand that the representatives of x and y at t' in that interval are unambiguously defined; that is, we want a very "local" version of injectivity as follows: in the situation of DNT3, respectively DNT4, the $t_1 \in T$, respectively t_1, t_2 may be chosen such that $z \in \Lambda_t$ is the unique element, such that $\varphi_{tt_1}(z) = z_1$, respectively x', y', z' in

$\Lambda_{t'}$ are unique, such that $\varphi_{t't}(z') = z, \varphi_{t't}(y') = y, \varphi_{t't}(x') = x$ or x, y, z are unique mapping to $\varphi_{tt'}(x), \varphi_{tt'}(y), \varphi_{tt'}(z)$, depending whether $t' < t$ or $t < t'$. Since we are able to take finite intersections of open intervals in the totally ordered set T, we may extend the foregoing observations to finite nontrivial chains $x_1 < x_2 < \ldots < x_n$ in Λ_t.

For $\lambda_t \in \Lambda_t$ we consider $L(\lambda_t) = \{a_{t'}, t' \leq t, \varphi_{t't}(a_{t'}) = \lambda_t\} \cup \{a_{t''}, t \leq t'', \varphi_{tt''}(\lambda_t) = a_{t''}\} = P(\lambda_t) \cup F(\lambda_t)$ (past and future of λ_t). $LT(\lambda_t) = \{t' \in T$ such that $\exists a_{t'} \in \Lambda_{t'}, a_{t'} \in L(\lambda_t)\}$.

As observed above, finite chains have a nontrivial lifetime (LT), but in general a directed system A_t in Λ_t need not "exist" in a T-interval, so in general the $C(\Lambda_t)$ with respect to the $\varphi^e_{tt'}, t \leq t'$ do not necessarily form a DNT.

Note also that "new" elements appear in some Λ_t as isolated elements; these may map under $\varphi_{tt'}, t \leq t'$, either to an isolated or an interacting element. The foregoing axioms do not contain information about the global life span of an element; for example, one might demand that "no element is an island," that is, remains isolated for all future $t' \geq t$; also one may demand that "no one lives forever," asking that for any $t, \lambda_t \in \Lambda_t$ there exists a $t' \geq t$ such that $\varphi_{tt'}(\lambda_t)$ is either 0 and 1. These assumptions are not relevant for our theory here; they do fit in a physics-oriented interpretation, though.

Definition 4.4: Observed Truth
Any statement depending on only finitely many ingredients of a DNT and depending on parametrization by $t \in T$ is said to be an **observed truth** at $t_0 \in T$ if there is an open interval $]t', t[$ in T containing t_0, depending on the finitely many ingredients, such that the statement holds for parameter values in this interval.

It seems that mathematical statements about the noncommutative DNT turn into "observed truth" when checked in the commutative shadows (meaning on the negative side that the same mathematical statements cannot be established globally in the commutative world, a situation perhaps deserving philosophical consideration).

Let us consider a DNT given by $\{\Lambda_t, t \in T\}$ with $\varphi_{tt'} : \Lambda_t \to \Lambda_{t'}$ for $t \leq t'$, and let us assume that all Λ_t are topologies of virtual opens, in particular the $\vee$ operations are all assumed to be commutative (we abbreviate this as DVT).

Proposition 4.4
Consider a DVT as above and take commutative shadows $SL(\Lambda_t)$ of Λ_t with maps $\varphi_{tt'} : SL(\Lambda_t) \to SL(\Lambda_{t'})$, for $t \leq t'$ in T, defined by restriction of $\varphi_{tt'}$ to idempotents of $\mathrm{id}_\wedge(\Lambda_t)$. The statement that $\{SL(\Lambda_t), t \in T\}$ is a DVT is also an observed truth.

Proof
As observed in Lemma 4.1, all $\varphi_{tt'}$ map $\wedge$-idempotents to $\wedge$-idempotents. DNT 1 is obvious. For DNT 2 we have to check that each $\varphi_{tt'}$ preserves the operation $\dot{\wedge}$. Look at $\varphi_{t_0t} : \Lambda_{t_0} \to \Lambda_t$ and $\sigma, \tau \in \mathrm{id}_\wedge(\Lambda_{t_0})$. □

If $\sigma < \tau$ (or conversely), then $\varphi_{t_0t}(\sigma) \leq \varphi_{t_0t}(\tau)$ and hence $\varphi_{t_0t}(\sigma) \dot{\wedge} \varphi_{t_0t}(\tau) = \varphi_{t_0t}(\sigma) = \varphi_{t_0t}(\sigma \dot{\wedge} \tau)$ (interchanging the role of σ and τ in the converse case). So in checking $\varphi_{t_0t}(\sigma)\dot{\wedge}\varphi_{t_0t}(\tau) = \varphi_{t_0t}(\sigma\dot{\wedge}\tau)$ we may assume σ and τ to be incomparable. Suppose the equality does not hold, that is, $\varphi_{t_0t}(\sigma\dot{\wedge}\tau) < \varphi_{t_0t}(\sigma)\dot{\wedge}\varphi_{t_0t}(\tau)$. Applying DNT5 allows the assumption $\varphi_{t_0t}(\sigma) \neq \varphi_{t_0t}(\tau)$. Now $\varphi_{t_0t}(\sigma)\dot{\wedge}\varphi_{t_0t}(\tau) = \varphi_{t_0t}(\sigma)$ (similar argument will hold with σ and τ interchanged) leads to $\varphi_{t_0t}(\sigma) \leq \varphi_{t_0t}(\tau)$, hence $\varphi_{t_0t}(\sigma) < \varphi_{t_0t}(\tau)$. Using DNT5 again, taking t close enough to t_0, we obtain $\sigma\dot{\wedge}\tau < \sigma_1 < \tau$ such that $\varphi_{t_0t}(\sigma\dot{\wedge}\tau) < \varphi_{t_0t}(\sigma_1) = \varphi_{t_0t}(\sigma) < \varphi_{t_0t}(\tau)$. The interval $[t_0, t]$ being small enough in order to have unambiguous representation, $\sigma_1 = \sigma$ follows, but that would contradict incomparability of σ and τ. Consequently, we have strict relations:

$$\varphi_{t_0t}(\sigma\dot{\wedge}\tau) < \varphi_{t_0t}(\sigma)\dot{\wedge}\varphi_{t_0t}(\tau) \quad < \varphi_{t_0t}(\sigma)$$
$$\varphi_{t_0t}(\sigma\dot{\wedge}\tau) < \varphi_{t_0t}(\sigma)\dot{\wedge}\varphi_{t_0t}(\tau) \quad < \varphi_{t_0t}(\tau)$$

In view of DNT3 we may assume that t is close enough to t_0 in order to have obtained $z \in \Lambda_{t_0}$ such that $\sigma\dot{\wedge}\tau < z < \sigma, \tau$ and $\varphi_{t_0t}(z) = \varphi_{t_0t}(\sigma)\dot{\wedge}\varphi_{t_0t}(\tau)$.

Were z not idempotent, then $\sigma\dot{\wedge}\tau < z \wedge z < \sigma$ would lead to $\varphi_{t_0t}(z \wedge z) = \varphi_{t_0t}(\sigma)\dot{\wedge}\varphi_{t_0t}(\tau)$ because φ_{t_0t} respects $\wedge$ and the latter is $\wedge$-idempotent in Λ_t, thus $\varphi_{t_0t}(z) = \varphi_{t_0t}(z \wedge z)$, but the unambiguity of representation guaranteed by the choice of t close enough to t_0 (DNT 5) then yields $z = z \wedge z$ or $z \in \mathrm{id}_\wedge(\Lambda)$. Then $z = \sigma\dot{\wedge}\tau$ by definition is a contradiction. Consequently, $\varphi_{t_0t}(\sigma\dot{\wedge}\tau) = \varphi_{t_0t}(\sigma)\dot{\wedge} \varphi_{t_0t}(\tau)$ holds for τ in some (small enough) T-interval containing t_0. This establishes that DNT 2 is an observed truth. To check DNT 3 for $\{SL(\Lambda_t), t \in T\}$ we start from $\sigma < \tau$ in $\mathrm{id}_\wedge(\Lambda_t)$, $t < t_1$ such that $z_1 \in \mathrm{id}_\wedge(\Lambda_{t_1})$ with $\varphi_{tt_1}(\sigma) < z_1 < \varphi_{tt_1}(\tau)$ exists. By DNT3 for $\{\Lambda_t, t \in T\}$ there is a $z \in \Lambda_t$, $\sigma < z < \tau$ such that $\varphi_{tt_1}(z) = z_1$. Using DNT5 for $\{\Lambda_t, t \in T\}$ again, as in the foregoing part of the proof, we arrive at $\varphi_{tt_1}(z) = z_1$ with z also $\wedge$-idempotent in Λ_t (for t_1 close enough to t). The proof of DNT4 for $\{SL(\Lambda_t), t \in T\}$ follows, in the same way; DNT5 for $\{SL(\Lambda_t), t \in T\}$ is equally obvious (unambiguity in a suitable T-interval allows to us to "pull back" idempotency). All the DNT axioms hold for $\{SL(\Lambda_t), t \in T\}$ in a suitable T-interval because we are dealing with only finitely many ingredients, so the statement that $\{SL(\Lambda_t), t \in T\}$ is a DVT is an observed truth. □

Let us now turn to the definition of **temporal points**. First we fix a notion of point; we have seen earlier in these notes that sometimes it is interesting to consider different notions simultaneously, for example, in terms of irreducibility, in terms of maximality of associated filters, in terms of a quantum basis. So we use from here on the notion point in any one of these senses but fixed throughout the sequel.

Consider a DVT $\{\Lambda_t, t \in T\}$ (typically each Λ_t would be the pattern topology of some noncommutative space $X_t, t \in T$). A $\lambda \in \Lambda_t$ is said to be a **temporal point** if there is an interval $]t_0, t_1[$ in T such that $t \in]t_0, t_1[$ and for some $t' \in]t_0, t_1[$ there is a point $p_{t'} \in \Lambda_{t'}$ such that either $t \leq t'$ and $\varphi_{tt'}(\lambda_t) = p_{t'}$, or $t' \leq t$ and $\varphi_{t't}(p_{t'}) = \lambda_t$. We say that λ_t is a **future point** if the foregoing holds with respect to an interval $[t, t_1[$.

The system $\{\Lambda_t, t \in T\}$ is said to be **temporally pointed** if for every $t \in T$, $\lambda_t \in \Lambda_t$ there exists a family of temporal points $\{p_{\alpha,t}, \alpha \in \mathcal{A}\}$ in Λ_t such that λ_t is covered by it, $\lambda_t = \vee\{p_{\alpha,t}, \alpha \in \mathcal{A}\}$ (note that for a virtual topology we assumed $\vee$-completeness so that $\vee$ over arbitrary families may be considered). Some notation: $T\mathcal{P}(\Lambda_t)$ is the set of temporal points of Λ_t, $P\mathcal{P}(\Lambda_t)$ and $F\mathcal{P}(\Lambda_t)$ are respectively the set of **past points** respectively **future points**, for example, $P\mathcal{P}(\Lambda_t) = \{\lambda_t \in \Lambda_t, \lambda_t = \varphi_{t't}(\gamma_{t'})$, where $t' \leq t$ and $\gamma_{t'} \in \mathcal{P}(\Lambda_{t'}\}$, the definition of $F\mathcal{P}(\Lambda_t)$ is similar. We write $T\text{Spec}(\Lambda_t)$ for the set of temporal points sometimes because by $\text{Spec}(\Lambda_t)$ we denote the set $\{p_{t'}$ point in $\Lambda_{t'}$, $p_{t'}$ defines a temporal point of $\Lambda_t\}$. For T-intervals $[t_1, t_2], [t_3, t_4]$ (similar for open intervals) we write $[t_1, t_2] < [t_3, t_4]$ if $t_1 \leq t_3$ and $t_2 \leq t_4$ and use $<$ as a partial order on T-intervals of the same type, that is, closed ones.

Definition 4.5: Space Continuum
This definition tries to build in certain continuous aspects without using functions or a group structure on T. A temporally pointed system $\{\Lambda_t, t \in T\}$ is said to be a **space continuum** if the following conditions hold:

SC.1 There is a minimal closed interval $I_t \ni t$ in T such that $T\text{Spec}(\Lambda_t)$ has support in I_t (note that we do not demand that $\text{Spec}(\Lambda_t)$ has support in I_t; that is, there may be many other points, in some $\Lambda_{t'}$ with t' far from t, representing temporal points in Λ_t). The set of points in $\Lambda_{t'}$ with $t' \in I_t$ is then called a **minimal spectrum** for $T\mathcal{P}(\Lambda_t)$, denoted by $\text{Spec}(\Lambda_t, I_t)$.

SC.2 For any open T-interval I such that $I_t \subset I$ there exists an open T-interval $I^*(t)$ with $t \in I^*(t)$ such that for all $t' \in I^*(t)$ we have $I_{t'} \subset I$. In other words, if a minimal spectrum for $T\mathcal{P}(\Lambda_t)$ is realized in an open T-interval, then in some open interval around t the minimal spectra of the spaces remain realized in that open T-interval.

SC.3 If $t \leq t'$ in T, then $I_t < I_{t'}$; this provides an "orientation" of the variation of the minimal spectra.

SC.4 Local preservation of directed sets. For given $t \leq t'$ in I_t and any directed set A_t in Λ_t, the subset $\{\gamma_t \in \mathcal{A}_t,$ there exists $\xi_t < \gamma_t$ in A_t such that $\varphi_{tt'}(\xi_t) < \varphi_{tt'}(\gamma_t)\}$ is cofinal in A_t. For $t'' \leq t$ in I_t there is a directed set $A_{t''}$ in $\Lambda_{t''}$ mapped by $\varphi_{t''t}$ to a cofinal subset of A_t.

In the foregoing we used "a" family of temporal points in the definition of a temporally pointed system, but it is clear that we may look at a canonically defined one, that is, "the" family of temporal points defining a temporally pointed system by taking all of these realized in the interval I_t.

A subset J of T is said to be open around $t \leq T$ if it is **relative open**, that is the intersection of I_t and an open T-interval. For an arbitrary $x \in \Pi\{\Lambda_t, t \in T\}$ we write $\sup(x) = \{t \in T, x_t \neq 0\}$ where we assume x is written as a series $(\ldots, x_t, \ldots)$ with $x_t \in \Lambda_t$ for $t \in T$. We have assumed that the Λ_t are suitable, nice noncommutative or virtual topologies, for example, pattern spaces of noncommutative topologies $X_t, t \in T$. There are therefore special elements in Λ_t that have been obtained from opens in X_t via constant directed systems, for example, $[\lambda_t]$ for $\lambda_t \in X_t$. An element $x = (\ldots, x_t, \ldots)$ is said to be **topologically accessible** if all x_t, $t \in \sup(x)$

are of the type $[\lambda_t]$; look back at the remarks preceding Lemma 2.13 and also look at Definition 2.2 and the role of the $[\lambda_t]$ in defining the point topology on the point spectrum $\mathrm{Sp}(\Lambda_t)$. An element x as above is said to be **t-accessible** if $\sup(x) = J$ is relative open around t and for all $t' \leq t''$ we have that $\varphi_{t't''}(x_{t'}) \leq x_{t''}$.

In case $I_t = \{t\}$, that is, Λ_t has enough points, then the points in an open for the point topology would be characterized by $\{p, p \leq [\lambda_t]\} = U(\lambda_t)$ for some $\lambda_t \in X_t$. When Λ_t does not have "enough" points (points do not constitute a quantum basis for Λ_t), then we have to modify the definition of point spectrum and point topology correspondingly. If x is t-accessible, say $x = (\ldots, x_t, \ldots)$ and $p_{t'} \in \mathrm{Spec}(\Lambda_t, I_t)$, then we write $p_{t'} \in x$ if $t' \in J$, where J is the relative open interval around t in the definition of x, and there exists an open T-interval $J_1 \subset J$ with $t' \in J_1$ such that for $t'' \in J_1$ we have $p_{t''} \leq x_{t''}$, where by $p_{t''}$ we denote $p_{t''} = \varphi_{t't''}(p_{t'})$ if $t' \leq t''$ or for $t'' \leq t'$ we let $p_{t''} \in \Lambda_{t''}$ be such that $\varphi_{t''t'}(p_{t''}) = p_{t'}$, that is, $\{p_{t''}, t'' \in J_1\}$ is the restriction of a temporal point representing $p_{t'}$ that is defined over a bigger interval $]t_0, t_1[$ containing both t and t'.

Define $U_t \subset \mathrm{Spec}(\Lambda_t, I_t)$ by putting $U_t = U_t(x)$ and $U_t(x) = \{p_{t'}, p_{t'} \in x \text{ for some } t' \in I_t\}$, where x is t-accessible in $\Pi\{\Lambda_t, t \in T\}$.

Theorem 4.2

The empty set together with the sets $U_t(x)$, x t-accessible, defines a topology or $\mathrm{Spec}(\Lambda_t, I_t)$.

Proof

Consider $x \neq y$ both t-accessible with respective T-intervals J, respectively J' contained in I_t. If $p_{t'} \in U_t(x) \cap U_t(y)$, then $t' \in J \cap J'$ and for every $t_1 \in J$, $p_{t_1} \leq x_{t_1}$ for every $t_2 \in J'$, $p_{t_2} \leq y_{t_2}$. Of course the interval $J \cap J'$ is relative open around t. If $t' \leq t''$ with $t'' \in J \cap J'$, then $p_{t''} = \varphi_{t't''}(p_{t'})$ is idempotent in $\Lambda_{t''}$ because $p_{t'}$ is idempotent in $\Lambda_{t'}$ as it is a point. Hence we obtain:

$$p_{t''} = p_{t''} \wedge p_{t''} \leq x_{t''} \wedge y_{t''}.$$

Obviously for all $t'' \leq t'''$ in $J \wedge J'$ we do have: $\varphi_{t''t'''}(x_{t''} \wedge y_{t''}) \leq x_{t'''} \wedge y_{t'''}$. On the other hand, for $t'' \leq t'$ we obtain: $\varphi_{t''t'}(p_{t''}) = p_{t'}$ and therefore $p_{t'} \leq \varphi_{t''t'}(x_{t''}) \leq x_{t'}$, as well as $p_{t'} \leq \varphi_{t''t'}(y_{t''}) \leq y_{t'}$. Hence, again by idempotency of $p_{t'}$ in $\Lambda_{t'}$, we arrive at $p_{t'} \leq x_{t'} \wedge y_{t'}$. By restricting $J \cap J'$ to the interval obtained by allowing only those $t'' \leq t'$ that belong to an (open) unambiguity interval for $p_{t'}$ we arrive at a relative open around t, say $J'' \subset J \cap J'$, containing t'. Now, for $p_{t''}$ with $t'' \in J''$ it follows that $p_{t''}$ is idempotent because both $p_{t''}$ and $p_{t''} \wedge p_{t''}$ map to $p_{t'}$ via $\varphi_{t''t'}$ for $t'' \leq t'$ (other t'' in J'' are no problem). Thus for t'' in J'' we do arrive at $p_{t''} \leq x_{t''} \wedge y_{t''}$. Define w by putting $w_{t''} = x_{t''} \wedge y_{t''}$ for $t'' \in J''$. Clearly w is t-accessible and $p_{t'} \in U_t(w)$. Conversely, if $p_{t'} \in U_t(w)$, then $p_{t'} \in U_t(x) \cap U_t(y)$ is clear because J'' used in the definition of w is open in $J \cap J'$. Now we look at a union of $U_{i,t} = U_t(x_i)$ for $i \in J$ and each x_i being t-accessible with corresponding relative open interval J_i in I_t. Define w over the "interval" $J = \cup_i\{J_i, i \in J\}$ by putting $w_t = \vee\{x_{i,t}, i \in J\}$ for $t \in J$. It is clear that J is relative open around t and for all $t_1 \leq t_2$ in J we have $\varphi_{t_1t_2}(w_{t_1}) \leq w_{t_2}$ because $\varphi_{t_1t_2}$ respects arbitrary $\vee$. Now $p_{t'} \in w$ means that

$p_{t''} \leq \vee\{x_{i,t''}, i \in \mathcal{I}\}$ for t'' in some relative open containing t', say $J_1 \subset J$. We use relative open sets in T because I_t was closed and there are two situations to consider concerning $t' \in I_t$. First if t' is the lowest element of I_t, then for all $t'' \in J_1$ we have that $p_{t''} = \varphi_{t't''}(p_{t'}) \leq \varphi_{t't''}(\vee\{x_{i,t'}, i \in J\})$ and for all $t' \leq t_1 \leq t''$ we also obtain $p_{t_1} \leq \varphi_{t't_1}(\vee\{x_{i,t'}, i \in J\})$ and $p_{t''} \leq \varphi_{t_1t''}(\vee\{x_{i,t_1} i \in J\})$.

Otherwise, if t' is not the lowest element of I_t, then we may restrict J_1 to be an open interval $]t_0, t_0'[$ containing t' with $t_0 \in J$. The same reasoning as in the first case yields for all $t'' \in]t_0, t_0'[$ that $p_{t''} \leq \varphi_{t_0t''}(\vee\{x_{i,t_0} i \in J\})$ and for any $t' \leq t_1 \leq t''$ $p_{t''} \leq \varphi_{t_1t''}(\vee\{x_{i,t_1}, i \in \mathcal{I}\}) = \vee\{\varphi_{t_1t''}(x_{i,t_1}), i \in \mathcal{I}\}$. Since $t' \in J_1$ we obtain $p_{t'} \leq \vee\{\varphi_{t_1t'}(x_{i,t_1}), i \in J\}$ for all $t_1 \in [t_0, t']$.

Since $p_{t'}$ is a point in $\Lambda_{t'}$ there is an $i_0 \in J$ such that $p_{t'} \leq \varphi_{t_0t'}(x_{i_0,t_0})$ and therefore we have that $p_{t'} \leq \varphi_{t_1t'}(x_{i_0,t_1})$ with $t_1 \in [t_0, t']$, the gain being that i_0 does not depend on t_1 here! Now for $t'' \geq t'$ in $J_1 \cap J_{i_0}$ (note that this is not empty because x_{i_0} is nonzero at t_0 because $p_{t'} \leq \varphi_{t_0t^n}(x_{i_0,t_0})$ would then make $p_{t'}$ zero and we do not look at the zero (the empty set) for the selected a point of Λ_t). We obtain:

$$(*) \qquad p_{t''} = \varphi_{t't''}(p_{t'}) \leq \varphi_{t't''}(x_{i_0,t'}) \leq x_{i_0,t''}$$

In the other situation $t'' \leq t'$ in $J_1 \cap J_{i_0}$ we have $\varphi_{t''t'}(p_{t''}) = p_{t'}, \varphi_{t''t'}(x_{i_0,t''}) \leq x_{i_0,t'}$. By restricting $J_1 \cap J_{i_0}$ further so that the $t'' \leq t'$ are only varying in an (open) unambiguity interval for $p_{t'}$, say $J_2 \subset J_1 \cap J_{i0}$, we arrive at one of two cases: either $p_{t''} = x_{i_0,t''}$ or else $p_{t''} \neq x_{i_0,t''}$ and also $p_{t'} < x_{i_0,t'}$. In the first case $p_{t'} \in x_{i_0}$ follows because $p_{t_1} = \varphi_{t''t_1}(x_{i_0,t''}) \leq x_{i_0,t_1}$ for t_1 in $]t'', 1[\cap J_2$, the latter interval containing t' is relative open again. In the second case we may look at $p_{t'} < x_{i_0,t'} < 1$; hence there exists a $z_{t''}$ such that $p_{t''} < z_{t''} < 1$ and $\varphi_{t''t'}(z_{t''}) = x_{i_0,t'}$. Again we have to distinguish two cases, first $\varphi_{t''t'}(x_{i_0,t''}) = x_{i_0,t'}$ or $\varphi_{t''t'}(x_{i_0,t''}) < x_{i_0,t'}$. In the first case $x_{t''}$ and $x_{i_0,t''}$ map to the same element via $\varphi_{t''t'}$; hence up to restricting the interval further such that t'' stays within in an unambiguity interval for $x_{i_0,t'}$, we may conclude $z_{t''} = x_{i_0,t''}$ in this case, and then $p_{t''} < x_{i_0,t''}$. In the second case we may look at:

$$p_{t'} \leq \varphi_{t''t'}(x_{i_0,t''}) < x_{i_0,t'} < 1$$

(where the first inequality stems from (*) above).
Again restricting the interval further (but open) we find a $z'_{t''}$ in $\Lambda_{t''}$ such that $x_{i_0,t''} < z'_{t''} < 1$ such that $\varphi_{t''t'}(z'_{t''}) = x_{i_0,t'}$. Since we are dealing with the case $p_{t''} \neq x_{i_0,t''}$ and we are in an unambiguity interval for $p_{t'}$ it follows that $p_{t'} < \varphi_{t''t'}(x_{i_0,t''})$. Look at $p_{t'} < \varphi_{t''t'}(x_{i_0,t''}) < x_{i_0,t'}$ with $\varphi_{t''t'}(p_{t''}) = p_{t'}$ and $\varphi_{t''t'}(z'_{t''}) = x_{i_0,t'}$; by restricting the interval (open) further if necessary we obtain the existence of $z''_{t''}$ such that, $p_{t''} < z''_{t''} < z'_{t''}$ such that $\varphi_{t''t'}(z''_{t''}) = \varphi_{t''t'}(x_{i_0,t''})$. Finally, restricting again the $t'' \leq t'$ to vary in an unambiguity interval for $\varphi_{t''t'}(x_{i_0,t''})$ it follows that $z''_{t''} = x_{i_0,t''}$ and hence $z''_{t''} \geq p_{t''}$ yields $x_{i_0,t''} \geq p_{t''}$ for t'' in a suitable relative open around t containing t'. Thus also in this case we arrive at $p_{t'} \in x_{i_0}$ or $p_{t'} \in U_t(x_{i_0})$. It follows that $U_t(w) = \cup\{U_{i,t}, i \in J\}$ establishing that arbitrary unions of opens are open. By taking $U_t(1)$ we obtain the whole spectrum at t as an open too. □

Now instead of looking at presheaves over the Λ_t connected by suitable morphisms lying over the $\varphi_{t't''}$, it is natural to look at (pre-)sheaf theory over $\text{Spec}(\Lambda_t, I_t)$ and its topology now defined in a classical way via sets. In particular, the sheafification construction should be considered in this context; for example, is it possible to construct an étale space of classical type, that is, depending on **sets of stalks**?

First observe that in the definition of DNT we did not demand the $\varphi_{tt'}$ be continuous in the gen-topology; note that this continuity appears as an observed truth in the sense that for a gen-open $O_{t'}$ in $\Lambda_{t'}$ and t close enough to t' we have $\varphi_{tt'}^{-1}(O_{t'})$ open in the gen-topology of Λ_t.

For now we fix a category $\underline{\mathcal{C}}$ allowing limits and colimits; for this section we may restrict attention to abelian categories or even Grothendieck categories for convenience's sake, but that is not essential. For every $t \in T$ we have given a presheaf Γ_t over Λ_t and for $t \leq t'$ in T there are $\phi_{tt'} : \Gamma_t \to \Gamma_{t'}$ defined by morphisms of $\underline{\mathcal{C}}$ as follows: for $\lambda_t \in \Lambda_t$ there is a $\phi_{tt'}(\lambda_t) : \Gamma_t(\lambda_t) \to \Gamma_{t'}(\varphi_{tt'}(\lambda_t))$ and for each $\mu_t \leq \lambda_t$ in Λ_t we have a commutative diagram in $\underline{\mathcal{C}}$:

$$(\Delta) \qquad \begin{array}{ccc} \Gamma_t(\lambda_t) & \xrightarrow[\phi_{tt'}(\lambda_t)]{} & \Gamma_{t'}(\varphi_{tt'}(\lambda_t)) \\ \downarrow{\scriptstyle \rho^t_{t,\mu_i}} & & \downarrow{\scriptstyle \rho^{t'}_{\lambda_{t'},\mu_{t'}}} \\ \Gamma_t(\mu_t) & \xrightarrow[\phi_{tt'}(\mu_t)]{} & \Gamma_{t'}(\varphi_{tt'}(\mu_t)) \end{array}$$

where we have written $\lambda_{t'}, \mu_{t'}$ respectively for $\varphi_{tt'}(\lambda_t), \varphi_{tt'}(\mu_t)$, and $\rho^t_{.,.}$ for the restriction morphisms in the presheaf Γ_t.

We demand, moreover, that ϕ_{tt} is defined by $\phi_{tt}(\lambda_t) = I_{\Gamma_t(\lambda_t)}$ and $\phi_{t't''}\phi_{tt'} = \phi_{tt''}$ for $t \leq t', t' \leq t''$, in the sense that for all $\lambda_t \in \Lambda_t$. $\phi_{t't''}(\phi_{tt'}(\Gamma_t(\lambda_{t''})) = \phi_{tt''}(\Gamma_t(\lambda_t))$. Under these conditions the system $\{\Gamma_t, \phi_{tt'}, t \leq t' \text{ in } T\}$ is called a **(global) dynamical presheaf** over the DNT $\{\Lambda_t, \varphi_{tt'}, t \leq t' \text{ in } T\}$. We can pass to stalks both in Γ_t and $\Gamma_{t'}$, but since the oriented sets defining these stalks need not be connected via $\varphi_{tt'}$ (because the T-interval containing t' where SC.4 would hold, need not contain t), we cannot relate these stalks in $\underline{\mathcal{C}}$ in an obvious way.

From here on we restrict to a temporally pointed system $\{\Lambda_t, t \in T\}$ satisfying the conditions of Definition 4.5. We look at Λ_t and Spec(Λ_t, I_t) with its topology given by open sets $U_t(x)$ associated to t-accessible elements x as in Theorem 4.2.

For $p_{t'} \in Y_t$ we may calculate $\Gamma_{t',p_{t'}} = \varinjlim \Gamma_{t'}(x_{t'})$ where $\varinjlim$ is over $x_{t'}$ such that $p_{t'} \leq x_{t'}$ x a t-accessible element; in fact we have $p_{t'} \in x$. Keep in mind that x is topologically accessible; that is, all $x_{t''}$, $t'' \in \sup(x)$ are of the type $[\lambda_{t''}]$ hence really representing an "open" of $\Lambda_{t''}$. In the foregoing we did not ask for the system $(\Lambda_{t'}, \varphi_{tt'}, t \leq t'$ in $T\}$ to derive from a system $\{X_t, t \in T\}$ with respect to some morphisms $X_t \to X_{t'}$ for $t \leq t'$ in T. Avoiding the introduction of the space continuum conditions on the level of the noncommutative topologies $X_t, t \in T$, we propose a minimal condition necessary to control t-accesibility up to a desired level; we prefer not to dwell upon the formal comparison of dynamical theories for the X_t and their associated pattern topologies Λ_t here.

Definition 4.6

An element $u_t \in \Lambda$ is said to be **classical** if it is of the form $[\lambda_t]$ for $\lambda_t \in X_t$; that is, the classical elements are the image of λ_t in Λ_t as in Lemma 2.1(3). If u_t is classical in Λ_t, then there is an open T-interval containing t, say L, such that for every $t' \in L$ we have that $u_{t'}$ is classical, where for $t \leq t'$, $u_{t'} = \varphi_{tt'}(u_t)$ and for $t' \leq t, u_{t'}$ is a chosen representative for u_t, $\varphi_{t't}(u_{t'})' = u_t$. Restricting further to an unambiguity interval of u_t the representatives $u_{t'}$ for t' in that interval are unique. If the foregoing condition holds, we say that the system $\{\Lambda_t, \varphi_{tt'}, t \leq t' \text{ in } T\}$ is **traditional**.

Lemma 4.2
For a traditional space continuum with dynamical presheaf $\{\Gamma_{t'}, \phi_{tt''}, t \leq t'' \text{ in } T\}$, *the stalk for* $p_{t'} \in Y_t$ *of* $\Gamma_{t'}$ *is exactly* $\Gamma_{t',p_{t'}}$ *as defined above.*

Proof
Since $p_{t'}$ is a point of $\Lambda_{t'}$, the stalk of the presheaf $\Gamma_{t'}$ is obtained by taking $\underset{p_{t'} \leq u_{t'}}{\overset{\lim}{\longrightarrow}} \Gamma_{t'}(u_{t'})$, the limit ranging over $u_{t'} \in \Lambda_{t'}$. Since points of $\Lambda_{t'}$ are defined as points of $X_{t'}$, the limit may be calculated over a cofinal system obtained by letting the $u_{t'}$ vary over classical elements of $\Lambda_{t'}$. The lemma follows if we establish existence of a t-accessible y such that $p_{t'} \in y$ and $y_{t'} \leq u_{t'}$. Now from $p_{t'} \in U_t(x)$ we obtain $(\ldots x_{t'}, \ldots)$ with a relative open T-interval of definition J say, $t' \in J$, such that $p_{t''} \leq x_{t''}$ for every $t'' \in J$. Since $u_{t'}$ and $x_{t'}$ are classical, so is $u_{t'} \wedge x_{t'}$ and moreover $p_{t'} \leq u_{t'} \wedge x_{t'}$ because $p_{t'}$ is a point (hence idempotent in $\Lambda_{t'}$). Let J_1 be an open T-interval containing t' such that $u_{t'} \wedge x_{t'}$ has a representative $u_{t''}$ in $\Lambda_{t''}$ such that $\varphi_{t''t'}(u_{t''}) = u_{t'} \wedge x_{t'}$. Since $x_{t'} \neq u_{t'} \wedge x_{t'}$ may be assumed (otherwise put $y = x$) we arrive at $x_{t''} > u_{t''} \geq p_{t''}$. Using the intersection of J_1 with the interval around t' allowing us to select classical $u_{t''}$, calling it J_2, we put $y_{t''} = u_{t''}$ for $t'' \leq t'$ in J_2 and $y_{t_1} = x_{t_1}$ for $t' < t_1$ in J. Then y is t-accessible in the relative open T-interval around t just defined, and we have $y_{t'} \leq u_{t'}$ and $p_{t'} \in y$. Consequently:

$$\underset{p_{t'} \geq u_{t'}}{\overset{\lim}{\longrightarrow}} \Gamma_{t'}(u_{t'}) = \underset{p_{t'} \in x}{\overset{\lim}{\longrightarrow}} \Gamma_{t'}(x_{t'}). \qquad \square$$

The former learns that we have unambiguously defined stalks at the elements of Y_t, so we aim to construct an étale space over Y_t by using the open sets $U_t(x)$, x t-accessible.

In the sequel we restrict attention to presheaves and sheaves with values in a category $\underline{C}$, the objects of which are at least sets. In fact, let us deal with the sheaf theory for $\underline{C}$, the category of abelian groups; the reader may translate the constructions either to **sets** or richer categories with objects that have an underlying group structure.

We consider again a traditional space continuum with dynamical presheaf $\{\Gamma_t, \phi_{tt'}, t \leq t' \text{ in } T\}$. On Y_t we now define a presheaf with respect to the topology as in Theorem 4.2 by taking for $\mathcal{P}(U_t(x))$ the abelian group in $\coprod_{t' \in I_t} \Gamma_{t'}(x_{t'})$ formed by the strings over $\sup(x) = \{t' \in I_t, x_{t'} \neq 0\}$, that is, $\{(\gamma_{t'}, t' \in \sup(x), \phi_{t''t'}(\gamma_{t''}) = \gamma_{t'} \text{ for all } t'' \leq t' \text{ in } \sup(x)\}$. Let us write $x < y$ if $x_{t'} \leq y_{t'}$ for all $t' \in I_t$; in particular $x < y$ forces $\sup(x) < \sup(y)$. For notational convenience we do not want to make a special case out of each $0 \in \Lambda_t$ so we agree to write $\Gamma_{t'}(0) = \underset{\longrightarrow}{\lim} \Gamma_{t'}(x_{t'})$, the limit being over all classical elements of $\Lambda_{t'}$ (the reader may prefer to restrict attention to nonzero (meaning "nonempty" in the classical sense) elements in a topology when dealing with presheaves and sheaves; this is easily decoded from our presentation).

If $x < y$, then $x_{t'} \leq y_{t'}$ for all $t' \in \sup(x)$ and we have restriction morphisms $\rho^{t'}_{y_{t'},x_{t'}}$, $\rho^{t'}_{y_{t'},x_{t'}} : \Gamma_{t'}(y_{t'}) \to \Gamma_{t'}(x_{t'})$. In view of the commutativity of the diagrams (Δ) before Definition 4.6 we obtain corresponding morphisms on strings over the respective supports: $\rho_{y,x} : \mathcal{P}(U_t(y)) \to \mathcal{P}(U_t(x))$.

For a point $p_{t'}$ let $\eta(p_{t'})$ be the set of $U_t(x)$ such that $p_{t'} \in U_t(x)$, that is, $p_{t'} \in x$; in particular $t' \in J_x$ where J_x is the relative open around t in the definition of x, thus $t' \subset \cap\{\sup(x), \eta(p_{t'}) \ni U_t(x)\}$.

$$\text{What is the stalk } \underset{U_t(x)\in\eta(\rho_{t'})}{\overset{\lim}{\longrightarrow}} \mathcal{P}(U_t(x))?$$

First, for the dynamical sheaf theory we may want to impose some property dual, in some sense, to the "very local" injectivity we assumed for the $\varphi_{tt'}$. It is natural to ask that $\Gamma_{t''}$ weakly approximates $\Gamma_{t''}$ for t'' close enough to t'.

Definition 4.7

The dynamical presheaf $\{\Gamma_{t'}, \phi_{t't''}, t' \leq t'' \text{ in } T\}$ on a traditional space continuum is **locally temporally flabby** at $t \in T$ if for t-accessible x, $t' \in \sup(x)$ such that $p_{t'} \in x$ and $s_{t'} \in \Gamma_{t'}(x_{t'})$, there exists a t-accessible $y < x$ with $p_{t'} \in y$ and a string $\overline{s} \in \mathcal{P}(U_t(y))$ such that $\overline{s}_{t'} = \rho^{t'}_{x_{t'},y_{t'}}(s_{t'})$.

Theorem 4.3

To a dynamical presheaf on a traditional space continuum there corresponds for every $t \in T$ *a presheaf* $\mathcal{P}_t$ *on the spectrum* $\mathrm{Spec}(\Lambda_t, I_t)$ *with its spectral topology given by the* $U_t(x)$, x *t-accessible. If all presheaves* $\Gamma_{t'}$ *for* $t' \in I_t$, *are separated, then* $\mathcal{P}_t$ *is separated. The sheafification* $\underline{\underline{a}}\mathcal{P}_t$ *of* $\mathcal{P}_t$ *on* $\mathrm{Spec}(\Lambda_t, I_t)$ *is called the* ***spectral sheaf at*** t. *In case the dynamical presheaf is locally temporally flabby, then for a point* $p_{t'} \in \mathrm{Spec}(\Lambda_t, I_t)$, *the stalk* $\mathcal{P}_{t,p_{t'}}$ *may be identified with the stalk* $\Gamma_{t',p_{t'}}$.

Proof

At every $t \in T$, $\mathcal{P}_t$ is the spectral presheaf constructed on $\mathrm{Spec}(\Lambda_t, I_t)$ with its spectral topology. Now suppose all Γ_t are separated presheaves, and look at a finite cover $U_t(x) = U_t(x_1) \cup \ldots \cup U_t(x_n)$ and a $\gamma \in \Gamma_t(U_t(x))$ such that for $i = 1, \ldots, n$ $\rho_{x,x_i}(\gamma) = 0$. We have seen before that the union $U_t(x_1) \cup \ldots \cup U_t(x_n)$ corresponds to the t-accessible element $x_1 \vee \ldots \vee x_n$ obtained as the string over $\sup(x_1) \cup \ldots \cup \sup(x_n)$ given by the $x_{1,t'} \vee \ldots \vee x_{n,t'}$ in $\Lambda_{t'}$. For all $t' \in \sup(x)$ we obtain, in view of the compatibility diagrams for restrictions and $\phi_{t't''}$, $t' \leq t''$: $\rho^{t'}_{x_{t'},x_{i,t'}}(\gamma_{t'}) = 0$, for $i = 1, \ldots, n$. The assumed separatednesss of $\Gamma_{t'}$, for all t', then leads to $\gamma_{t'} = 0$ for all $t' \in \sup(x)$ and therefore $\gamma = 0$ as a string over $\sup(x)$. Consequently $\mathcal{P}_t$ is separated, for all $t \in T$. In order to calculate the stalk at $p_{t'} \in \mathrm{Spect}(\Lambda_t, I_t)$ for $\mathcal{P}_t$ we have to calculate $\underset{p_{t'}\in x}{\overset{\lim}{\longrightarrow}} \mathcal{P}_t(U_t(x)) = E_{t'}$. Starting with $p_{t'} \in x$ for some t-accessible x we have a representative $\gamma_x \in \mathcal{P}_t(U_t(x))$ being a string over sup(x) and the latter containing a relative open $J(x)$ around t containing t'. So an element $e_{t'}$ in $E_{t'}$ may be viewed as given by a direct family $\{\gamma_x, p_{t'} \in x, \rho_{x,y}(\gamma_x) = \gamma_y$ for $y < x\}$. At t', which is in sup(x) for all x appearing in the foregoing family (as $U_t(x)$ varies over $\eta(p_{t'})$), we obtain $\{(\gamma_{x'})_{t'}, p_{t'} \le x_{t'}, \rho^{t'}_{x_{t'},y_{t'}}(((\gamma_x)_{t'}) = (\gamma_y)_{t'}\}$, which defines an element of $\Gamma_{t',p_{t'}}$, say $\overline{e}_{t'}$. We have a well-defined map $\pi_{t'} : E_{t'} \to \Gamma_{t',p_{t'}}, e_{t'} \mapsto \overline{e}_{t'}$. Without further assumption we therefore arrive at a sheaf $\underline{\underline{a}}\mathcal{P}_t$ with stalk $E_{t'}$ at $p_{t'}$ and a presheaf map $\mathcal{P}_t \to \underline{\underline{a}}\mathcal{P}_t$, which is "injective" if all $\Gamma_{t''}$ are separated. Now we have to make use of the local temporally flabbyness (LTF). Look at a germ $s_{t'} \in (\Gamma_{t'})_{p_{t'}}$. In view of Lemma 4.2 there exists a t-accessible x such that $s_{t'} \in \Gamma_{t'}(x_{t'})$ with $p_{t'} \in x$, in particular $p_{t'} \le x_{t'}$. The LTF condition allows us to select a t-accessible $y < x$ with $p_{t'} \in y$ together with a string, $\vec{s}(y) \in \mathcal{P}(U_t(y))$ such that $\vec{s}_{t'}(y) = \rho^{t'}_{x_{t'},y_{t'}}(s_{t'})$. The element $e_{t'}$ in $E_{t'}$ defined by the directed family obtained by taking restrictions of $\vec{s}_{t'}(y)$ has $\overline{e}_{t'}$ exactly $s_{t'}$ (note that t' supports all the restrictions of $\vec{s}_{t'}(y)$ because y varies in $\eta(p_{t'})$). Thus $\pi_{t'} : E_{t'} \to \Gamma_{t',p_{t'}}$ is epimorphic. If $e_{t'}$ and $e'_{t'}$ have the same image under $\pi_{t'}$ then there is a t-accessible y such that $e_{t'} - e'_{t'}$ is represented by the zero-string over sup(y); in fact this follows by taking $s_{t'} = 0$ in the foregoing leading to a t-accessible y as above, which we may restrict to a t-accessible y' defined by taking for sup(y') the relative open J containing t' in the support of y where $\vec{s}_{t'}(y) = 0$, and putting $y_{t''}$ for $t'' \in$ sup(y'). Therefore, $\pi_{t'}$ is also injective. □

Remark 4.1

Can one conclude the injectivity of $\pi_{t'}$, at least, without the LTF condition? It seems that the idea of *germ* is needed in the temporal direction to complement the topological germs appearing in stalks, so there may be other phrasings of extra axioms, but probably the LTF condition is, in some form, unavoidable.

4.2.2 Project: Spectral Families on the Spectrum

In Section 2.7 we defined spectral families and the corresponding observables on a noncommutative topology in relation to the generalized Stone space. Now we may define in exactly the same way Γ-spectral families on the spectrum Spec$(\Lambda_{t'}, I_t)$ in terms of its spectral topology. In other words we look at suitable Γ-indexed families $\{U_t(x_\gamma), \gamma \in \Gamma\}$, where each x_j is t-accessible, defining a separated Γ-filtration. For $t'' \in I_t$ we may look at $V_t(x_\gamma) = \{p_{t''}$ where $p_{t'} \in U_t(x_\gamma)\}$ again $p_{t''} = \varphi_{t't''}(p_{t'})$ or $\varphi_{t't''}(p_{t'}) = p_{t''}$ depending upon whether $t' \le t''$ or $t'' \le t'$. The family $\{V_t(x_\gamma), \gamma \in \Gamma\}$ need not be a spectral family at moment t''. The project exists in describing transitions of spectral families via the system $\varphi_{t't''}$ depending on the condition of local preservation of directed systems. A stronger notion of spectral family at $t \in T$ may be obtained by demanding existence of spectral families "stringwise" in a relative open J around t, in the straightforward way. This stronger definition indeed implies the existence of a spectral family at t and moreover on $\Lambda_{t''}, t'' \in J$, but not necessarily on Spec$(\Lambda_{t''}, I_{t''})$for $t'' \in J$! The problem here is as always the (lack of) relation between I_t and $I_{t''}$. Relate the different definitions possible. In case T is also a group or whenever intervals may be "measured" or compared in "length" via translations, for example, the behavior of spectral families in time may become easier to describe.

4.2.3 Project: Temporal Čech and Sheaf Cohomology

In [50] the authors developed Čech cohomology on the noncommutative site and used it to calculate sheaf cohomology. Now we may want to derive the T-string version of this in order to arrive at relations between Čech cohomology for the noncommutative topology Λ_t versus the one for the spectral topology at t. In this theory the notion of affine elements (see Chapter 3) may play a role, but this in turn would require a T-string version of Grothendieck representations.

4.2.3.1 Subproject 1: Temporal Grothendieck Representations

A general theory may be developed by starting from different Grothendieck representations for each Λ_t, that is, different families of Grothendieck representations related by suitable connecting families of functors defined over the $\varphi_{t't''}$. Stringwise defined *affineness* should then be compared to affineness at t'' in a suitable relative open (around t). A good example may then be developed by looking at the special case of essential functors and structure sheaves.

A less general approach would perhaps fit the "temporal" philosophy better; indeed we adapted the latter in order to have at each moment t enough points available in the geometry at t and have a spectrum, a set theoretic topology, and a corresponding sheaf theory, in particular a sheafification technique. So logically a Grothendieck representation may be viewed as a single family of Grothendieck categories $\underline{G}$ but with (different) representations $\mathrm{Rep}_t : \Lambda_t \to \underline{G}$ connected over the $\varphi_{t't''}$, $t' \leq t''$ in T by natural transforms $c_{t't''}$ defined as a coherent system over the diagrams; for $t \leq t''$ in T:

$$\begin{array}{ccc} \lambda_{t'} & \xrightarrow{\varphi_{t't''}} & \lambda_{t''} \\ \uparrow & & \uparrow \\ \mu_{t'} & \longrightarrow & \mu_{t''} \end{array} \qquad \begin{array}{ccc} \mathrm{Rep}_{t'}(\lambda_{t'}) & \longleftarrow & \mathrm{Rep}_{t''}(\lambda_{t''}) \\ \uparrow & & \uparrow \\ \mathrm{Rep}_{t'}(\mu_{t'}) & \longleftarrow & \mathrm{Rep}_{t''}(\mu_{t''}) \end{array}$$

where $\mu_{t'} \leq \lambda_{t'}$, $\mu_{t''} \leq \lambda_{t''}$, $\varphi_{t't''}(\lambda_{t'}) = \lambda_{t''}$, $\varphi_{t't''}(\mu_{t'}) = \mu_{t''}$.

In this situation preservation of affineness is easier to describe and conditions for this to occur may be more natural. Also when applying this to the functorial geometry, for example, in terms of essential functors or localization functors, the use of natural transformations is indeed natural!

4.2.3.2 Subproject 2: Temporal Čech Cohomology and Sheaf Cohomology

For schematic algebras the use of a noncommutative topology based (literally) on Ore sets automatically puts us in the situation where the noncommutative topology has a basis of affine sets. The earlier project of extending this to abstract noncommutative topologies includes the manipulation of affine sets as in Chapter 3, imposing conditions on the topology concerning the existence of a basis of affine elements if necessary (schematic topologies). The foregoing subproject would lead to temporal versions of this. In both cases, that is, varying families of Grothendieck categories and representations or one fixed family of Grothendieck categories but varying representations suitably connected by natural transforms, a T-string version of Čech cohomology is easily written down. The first obvious problem again resides in relating this string version around $t \in T''$ to the actual cohomology at t (i.e., the one for Λ_t); the second problem is the more interesting relation between the string version of Čech cohomology and the classical Čech cohomology for the spectral topology on $\mathrm{Spec}(\Lambda_t, I_t)$.

It seems that noncommutative facts again transform to "observed truth" statements locally at $t \in T$, at least in the absence of extra conditions allowing us to control the t-intervals over which certain phenomena appear; so the final result seems to be a new kind of *observed Čech cohomology*. In case all presheaves Γ_t are in fact sheaves, we know that $\mathcal{P}_t$ is certainly a separated presheaf but it need not be a sheaf. It is an *observed sheaf* in the sense that the gluing axiom becomes an observed truth. The relation between Čech cohomology over Λ_t and sheaf cohomology does not carry over perfectly via its T-string version to the spectral topology and the sheaf cohomology over it. Is the spectral version at t, such as classical Čech cohomology on the spectral topology calculating sheaf cohomology, obtainable from the T-string version at least in the case where all Λ_t are schematic?

4.2.4 Project: Dynamical Grothendieck Topologies

It is not hard to start from a family of categories $\{\underline{\mathcal{C}}_t; t \in T\}$ with functors $\Phi_{tt'} : \underline{\mathcal{C}}_t \to \underline{\mathcal{C}}_{t'}$ for $t \leq t'$ in a totally ordered set T and identify the properties of the $\Phi_{tt'}$ in order to preserve covers and the pullback conditions; see Definition 2.5. A serious modification in the dynamical theory expounded before seems to be necessary in order to obtain some classical Grothendieck topological at moment t from string-wise constructions in the parametrized family of noncommutative ones. This may be a project of independent interest but since there is no relation with point-notions here we give it no priority for now. On the other hand, in Theorem 2.2 we established that Λ, $C(\Lambda)$, τ, the skew topologies we usually considered before, together with generic relations and covers induced by global covers, are examples of noncommutative Grothendieck topologies. Let X be the noncommutative Grothendieck topology thus obtained from Λ (or from $C(\Lambda)$, τ, ...) via $\underline{\mathcal{C}}^g$ as in Theorem 2.2. Since the generic quality of some relation in $\underline{\mathcal{C}}_t^g$ is visible on finite bracketed expressions, restriction to the intersection of finitely many unambiguity intervals of the "opens" where the patterns describing the generic property of the given relation are being evaluated, allows us to conclude that a generic relation is also alive on some open T-interval containing t. Obviously, a global cover lives on an open T-interval too, and a cover induced in an element by some global cover gives right to the string version of this property over a T-interval of unambiguity obtained by intersecting the corresponding unambiguity intervals of the element considered and the elements in the global covers.

In particular, it follows from DNT3 that if the relations $\varphi_{tt_1})(x) < z_1 < \varphi_{tt_1}(y)$ are generic, the $z \in \Lambda_t$ obtained may be chosen such that the relations $x < z < y$ are generic. Similarly in DTT4 the interval $]t_1, t_2[$ may be chosen small enough so that for generic $x < z < y$ in $\wedge_t$ the $\varphi_{tt'}(x) < \varphi_{tt'}(z) < \varphi_{tt'}(y)$ if $t' \leq t$ or the $x' < z' < y'$ with $\varphi_{t't}(z') = z$ if $t \leq t'$, are both generic. Prove the following lemma and study presheaves and sheaves on the dynamic Grothendieck topologies (GT) induced by the dynamical (pre)sheaves over the skew topologies.

Lemma 4.3
The functors $B_{tt'} : \underline{\mathcal{C}}_t \to \underline{\mathcal{C}}_{t'}$ induced by the $\varphi_{tt'} : \Lambda_t \to \Lambda_{t'}$ form an observed dynamical GT, that is the statement that it is a dynamical GT is an observed truth.

Remark 4.2
If a noncommutative topology Λ appears in a dynamical setup as described before, say $\Lambda = \Lambda_t$ for some fixed $t \in T$, then there is a sheafification procedure for (separated) presheaves Γ in $\mathcal{Q}(\Lambda, \underline{\mathcal{C}})$, where $\underline{\mathcal{C}}$ is an arbitrary Grothendieck category, appearing at $t \in T$ for some dynamical presheaf $\{\Gamma_t, \phi_{[t+1}, t \leq t' \text{ in } T\}$. The sheafification given by $\underline{\underline{a}}\mathcal{Q}_t$ as in Theorem 4.3 is defined on the spectral space that is a commutative topology. On the other hand, separated presheaves over Λ form the torsion-free class corresponding to a torsion theory on the Grothendieck category $\mathcal{Q}(\Lambda, \underline{\mathcal{C}})$ of presheaves over Λ with values in $\underline{\mathcal{C}}$, so we obtain a localization functor $\underline{\underline{s}} : \mathcal{Q}(\Lambda, \underline{\mathcal{C}}) \to \mathcal{S}(\Lambda, \underline{\mathcal{C}})$ and we may think of $\mathcal{S}(\Lambda, \underline{\mathcal{C}})$ as the category of *categorical sheaves* on Λ. The latter are now not related to suitable étale spaces via stalks at points, but of course if Λ is commutative, that is, a usual topology on a set of points, then the categorical sheaves are exactly the usual sheaves. Note that a dynamical (separable) presheaf $\{\Gamma_t, \phi_{tt'}, t \leq t'\}$ does yield a dynamical presheaf $\{s(\Gamma_t), \phi^s_{tt'}, t \leq t'\}$ where $\phi^s_{tt'} : s(\Gamma_t) \to s(\Gamma_{t'})$ is the sheaf morphism deriving from $\phi_{tt'} : \Gamma_t \to \Gamma_{t'}$ via obvious limit constructions in the category $\underline{\mathcal{C}}$. The sheaf $\underline{\underline{a}}\mathcal{Q}_t$ embeds in $\underline{\underline{a}}\mathcal{Q}_t(\underline{\underline{a}}(\Gamma_t))$ but need not be equal to it in general; that is, the categorical sheafification need not be compatible with the spectral sheafification.

4.2.5 Conjecture

If $\underline{\mathcal{C}}$ is the Rep-category defining the canonical topology Λ, then $\underline{\underline{a}}\mathcal{Q} = \underline{\underline{a}}\mathcal{Q}(\underline{\underline{s}}(\Gamma))$ holds. Variations on the conditions relating essential extensions in $\underline{\mathcal{C}}$ to the constructions over Λ and over $\text{Spec}(\Lambda_t, I_t)$ necessary in both sheafifications can be further investigated.

That $\underline{\underline{a}}$ and $\underline{\underline{s}}$ may not be tightly related in general should not be surprising. Indeed $\underline{\underline{s}}$ fits completely in a topos theoretic approach and $\underline{\underline{a}}$ reflects exactly the fact that sheaves on a noncommutative topology do not form a topos since it is constructed over a different space. Even though the "dynamical" theory introduced in Section 4.2 provides, on the pure mathematical level, many new ideas and interesting problems concerning new phenomena, it is clear that it does not provide an effective method of sheafification. Indeed, given a separated presheaf Γ on a noncommutative topology Λ, how can one create the blow-up blow-down construction effectively? There seem to be so many degrees of freedom in choosing the dynamic system $\{\Lambda_t, \varphi_{tt'}, t \leq t' \text{ in } T\}$ and corresponding global presheaf $\{\Gamma_t, \phi_{tt'}, t \leq t' \text{ in } T\}$ such that $\Lambda_{t_0} = \Lambda$, $\Gamma_{t_0} = \Gamma$ for some $t_0 \in T$, but in order to satisfy the axiom that there are "enough points in the universe" one has to define the $\varphi_{tt'}$ such that points are appearing at suitable $t \in T$ so that enough temporal points are available at each Λ_t. It quickly shows that one needs to understand much more about creation of points via the deformation-like maps $\varphi_{tt'}$; hence creating such a universe is difficult!

4.3 The Spaced-Time Model

Why noncommutative geometry if it cannot be related to reality in some sense? Trying to describe the unknown reality that exists outside us (?) by use of the mathematical formalism inside us is close to doing physics and not too far from philosophy as well. So perhaps this short section should be understood as part of mathematical physics at a foundational level. In the first part we describe a special case of the construction in Section 4.2, in fact this may be seen as looking at possible definitions for **noncommutative manifolds**; in the second part we propose a noncommutative space model for reality and indulge in some physical interpretations of some consequences of the noncommutative space model. Therefore, Section 4.3.2 is of an almost philosophical nature, containing some reactions resulting from my recent contacts with physicists; it should mainly be viewed as food for further thought.

4.3.1 Noncommutative Manifolds

In our dynamical mathematical theory of Section 4.2 all Λ_t may be different and also there is no reason to aim at invariance under t-variation for the commutative shadow $SL(\Lambda_t)$ or $\text{Spec}(\Lambda_t.I_t)$. From a more physics-oriented point of view, only one uniquely defined special case is very important, namely "reality" or a good (= accepted) model for it. From this point of view it is reasonable to assume that $SL(\Lambda_t)$ is the abstract geometric frame we reason in about reality, that is, 4-dimensional real space, space-time, or at least some suitable manifold. Then, as a first approximation, it is also reasonable to assume that all $SL(\Lambda_t)$ are isomorphic and that the maps $\varphi_{tt'}, t \leq t'$ in T, define the analytical isomorphisms $SL(\Lambda_t) \to SL(\Lambda_{t'})$. Now it was an observed truth that the system $(SL(\Lambda_t), \overline{\varphi}_{tt'}, t \leq t' \text{ in } T)$ is again a dynamical topology, where $\overline{\varphi}_{tt'}$ denoted the restriction of $\varphi_{tt'}$ to the idempotent elements (see Proposition 4.4). Also from the general theory, even if all $SL(\Lambda_t)$ are identical, it does not follow that the spectral spaces (sometimes we refer to these as *moment space*) $\text{Spec}(\Lambda_t, \Gamma_t)$ are closely related, let alone identical. In order to make calculations in mathematical structures not changing "during" the calculations it is acceptable to start from a model where $\text{Spec}(\Lambda_t, I_t)$ is a fixed commutative (topology of a) space so that two commutative objects (manifolds) and the evolution maps of the geometry $\varphi_{tt'}, t \leq t'$ in T, induce manifold isomorphisms $\underline{\varphi}_{tt'}$ on $SL(\Lambda_t)$. Note that (see second remark after Theorem 4.2) it is an observed truth that the maps $\varphi_{tt'}$ are continuous in the gen-topology of the $\Lambda_t, t \in T$. Let Y be (the Stone topology of) a fixed commutative space (manifold) isomorphic to the $SL(\Lambda_t)$; that is we assume given isomorphisms $\gamma_t : Y \to SL(\Lambda_t)$ such that $\varphi_{tt'}\gamma_t = \gamma_{t'}$ for all $t \leq t'$ in T. Similarly let X be (the Stone topology of) a fixed commutative space (manifold) isomorphic to the $\text{Spec}(\Lambda_t, I_t)$; that is, we assume given structral isomorphisms $\sigma_t : X \to \text{Spec}(\Lambda_t, I_t)$. Where Y is the framework for abstract reasoning about the noncommutative reality, that is, using an interpetation in space-time for example, the space X deals with "observed" phenomena. Typically X will be a higher dimensional space when compared to Y because we have blown up Y over T, which appears as a kind of irreversible time, for the moment only assumed to be a totally ordered set that may be thought of as having a rank; for example, one could look at $(\mathbb{R}_+)^n$, where $\mathbb{R}_+$ are the positive real numbers, and say it has rank n. Such augmentation in dimension would express that *measuring takes time and observation "creates space" out of time*! To analyze mathematical observations or measurements we reason in X. The uncertainty principle relating the nonsimultaneousness of observations to noncommutativity of operators or in fact of underlying space, if one thinks further, is now inherent in the mathematical formalism of the dynamical noncommutative space governed by the evolution maps $\varphi_{tt'}$ that define and are defined by the change in momentaneous geometry. Passing from Y to the noncommutative dynamical geometry is a version of deformation quantization but passing then further to X is to be viewed as a "dequantization" to a new commutative space allowing us to reason in terms of classical (commutative) mathematics about observations of reality.

An observed point in $\text{Spec}(\Lambda_t, I_t)$ is given by a string of elements, say $p_{t'} \in \Lambda_{t'}, t' \in J \subset I_t$ with $p_t \in \Lambda_t$ a temporal point. If $p_{t'}$ is a point hence idempotent in the noncommutative topology, then for all $t' \leq t''$, $\varphi_{t't''}(p_{t'})$ is idempotent and as such it appears in the commutative shadow $SL(\Lambda_{t''})$. Consequently, an observed point in $\text{Spec}(\Lambda_t, I_t)$ appears as a string in the base space Y; however, the string in Y may start later than t when the point actually existed already in noncommutative space. On the other hand, every $p_{t'} \in \Lambda_{t'}$ as above, point or not, is decomposable as a union of temporal points of $\Lambda_{t'}$ because the system $\{\Lambda_t, \varphi_{tt'}, t \leq t' \text{ in } T\}$ is temporally pointed. Consequently, every $p_{t'}$ is realized as an open $u(p_{t'})$ of $\text{Spec}(\Lambda_{t'}, I_{t'})$; thus in X the observed point appears as a string connecting opens, that is, a higher-dimensional type of string that could be thought of as a higher-dimensional tube (brane?). Note that we do not need the fact that a map $s_{tt'}$ from $\text{Spec}(\Lambda_t, t)$ to $\text{Spec}(\Lambda_{t'}, I_{t'})$ exists as an observed truth here because the given strings at the Λ_t-level do define sequences of elements (opens) in $\text{Spec}(\Lambda_t, I_t)$. The above describes the spaced-time model, the terminology obviously describing the varying noncommutative geometries indexed by an irreversible "time" parameter, having space-time for its commutative shadow Y and a kind of "brane space" for its dequantization X. Now it is possible to provide an approximate picture of *noncommutative space* as this model describes it.

Dynamical Noncommutative Geometry

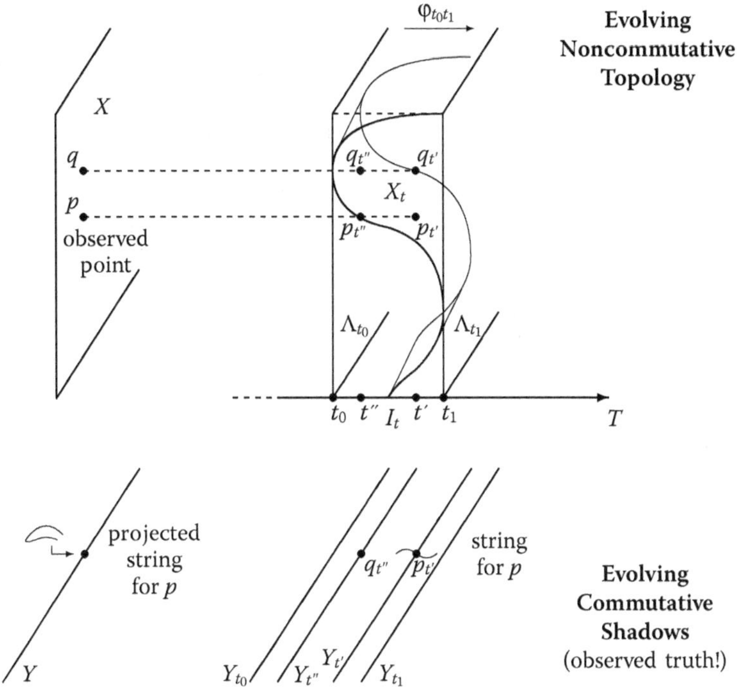

where we fixed (analytical) isomorphisms

$$\gamma_t : Y \to Y_t$$
$$\sigma_t : X \to X_t$$

The $\gamma_{t'}^{-1}(p_{t'})$ form a string in Y. The $\sigma_{t'}^{-1}(U(p_{t'})$ form a higher-dimensional string in X.

Below is an approximating visialization of these objects:

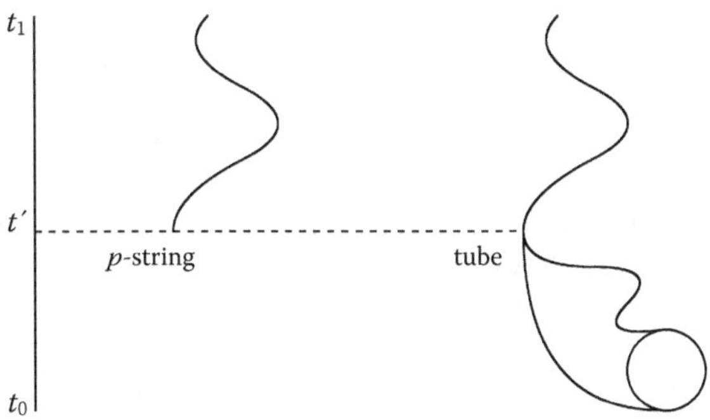

Observe that any temporal point defines points of $\Lambda_{t''}$ for all t'' satisfying $t_0 \leq t''$ for some $t_0 \in T$. Hence, "at infinite time" or rather "at the end of time" if you allow this somewhat loaded language, all temporal points have become points and the geometry at ∞'' is a commutative geometry, hence coinciding with its commutative shadow. This reminds us of the situation of almost commutative filtered algebras (i.e., having a commutative associated graded ring), with the Weyl algebras as the most famous examples, where the projective space of the associated graded ring appears as the space at infinity for the projective space obtained from the Rees ring (blow-up ring) of the filtration. It may again be just a parallel, but the dynamical topology seems to behave as a "blow-up" construction that is quantun-commutative in the sense that it is a deformation of a commutative space, which also appears as a (kind of) limit. The geometry of X should then be thought of as a geometrization related to $Y \times T$ up to some deformation.

So far, we did not relate T to the time parameter imbedded in Y (maybe we should call that relativistic time?), but it is possible to impose suitable mild connections, for example, the existence of order maps $\tau_t : T \to Y$ identifying $\tau_t(T)$ to a one-dimensional linear subspace of Y so that $\varphi_{tt'}\tau_t(t_0) = \tau_{t'}(t_0) + c_{t'}$ for all $t_0 \in T$, where $c_{t'}$ independent of t_0 is a constant in $\tau_{t'}(T)$ describing the translation of time observed in space-time for the transition from t to t' in T. We do not go deeper into this kind of modification trying to fit the model closer to presupposed reality. Nevertheless, let us mention some possible consequences of accepting the noncommutative world in the next section.

Another point we avoided is the introduction of real (or complex) numbers, for example, via manifolds or coordinates. We can reintroduce the reals if so desired.

4.3.1.1 Toward Real Noncommutative Manifolds

In a sense, the situation described above could be taken as a definition of a noncommutative manifold having Y for its commutative shadow; it is possible to be somewhat more specific in defining the real space structure. Start from a dynamic system ($\Lambda_t, \varphi_{tt'}, t \leq t'$ in T) with X and Y as before, X being a real manifold. We say that this system defines a **noncommutative manifold** if the following conditions hold. Consider a temporal point λ_t in Λ_t, say a future point, that is, $p_{t'} = \varphi_{tt'}(\lambda_t)$ is a point in $\Lambda_{t'}$; then there is a $\mu_t \geq \lambda_t$ such that $\varphi_{tt'}(\mu_t)$ is idempotent in $\Lambda_{t'}$ and $\varphi_{tt'}(\mu_t)$ is idempotent in $\Lambda_{t'}$ and $\varphi_{tt'}(\mu_t) \cong U_{t'}$ open in Y where $U_{t'}$ is a traditional open in the Stone topology of the topology of Y and $\cong$ meaning that the interval $[\circ, U_{t'}]$ in Y is isomorphic to $[\circ, \varphi_{tt'}(\mu_{t'})]$ in $SL(\Lambda_{t'})$ via $\gamma_{t'}$. Since $\overline{\varphi}_{tt'}$ is assumed to be an isomorphism for all $t \leq t'$ in T, there exists $\overline{\varphi}^{1}_{tt'}(\varphi_{tt'}(\mu_t)) = \alpha_t$, which is an idempotent in Λ_t; we demand that $\alpha_t \cong V_t$ open in Y where $\cong$ is understood as explained above. Observe that both μ_t and α_t map to $\varphi_{tt'}(\mu_t)$ but need not be equal because t' may be located outside the unambiguity interval for μ_t or α_t (note that equality would follow if μ_t is also idempotent because $\overline{\varphi}_{tt'}$ is assumed to be an isomorphism). A similar condition is imposed on past points but since such temporal points are also idempotent, the assumptions on $\overline{\varphi}_{tt'}$ entail extra unambiguity in the construction. Note that the temporal point λ_t in Λ_t gives rise to points $p_{t'}$ in $\Lambda_{t'}$ and $\overline{\varphi}^{-1}_{tt'}(p_{t'})$ in Λ_t together with opens respectively in $\Lambda_{t'}$ and Λ_t given by $p_{t'} \leq \varphi_{tt'}(\mu_t)$ respectively $\overline{\varphi}^{-1}_{tt'}(p_{t'}) \leq \alpha_t$, describing a local real structure around these points. Keep in mind that we have considered Y

to be the Stone topology of the topology of the manifold. This is only necessary because we have chosen to look at dynamical systems of Λ_t appearing as $C(X_t)$ for some X_t, a noncommutative topology, and restricting attention to traditional systems. This is therefore just a technical matter of presentation; for the sake of a physical model we could just have worked with a system of X_t and corresponding X and Y as defined before.

4.3.2 Food for Thought: From Physics to Philosophy

Let us accept for the moment that the spaced time model, up to possible further adaptions related to local Euclidean properties or connections between T and time as in space-time, is a good mathematical model for describing natural phenomena. What are the consequences of this? Well, the whole physics has to be reinterpreted because even if we accept the mathematical formalism phrased in space-time as a good approximation, its interpretation as a "projection" of a noncommutative reality allows many new interpretations knowing that some properties appear as observed truth (cannot be contradicted in some spatial and temporal neighborhood without being globally verifiable). At large intervals in space or time, the geometry itself is too different and physical entities have to be reinterpreted or regauged, even the notion of "distance" or "metric" for example. A possible example could be found in the recently discovered supernova that seemed to contradict the belief that the universe has positive curvature because calculations revealed it happened too close to us, however the (commutative space) distance is probably large enough so that the time interval necessary to measure it is larger than the unambiguity intervals of elements involved in the calculations, so that the commutative geometry at moment t_0 does not yield a good approximation of the reality at moment t in Λ_t (in other words the real distance should be calculated via noncommutative space if we could!) The ancient Greek idea of putting "ratios" at the center of geometry did not vanish with the discovery of $\sqrt{2}$ not being rational. The same thought extended to real numbers underlies the idea that some physical entity is expressible as a real number times a so-called unit, and concrete measurements are then expressible in the units one defines. However, there is no reason for two measurements of some entity to be even comparable, let alone to be a real multiple of one another. Moreover, since measurements take time, the noncommutative model shows that an uncertainty principle is already present at the level of mathematical description! Is the evolution of our awareness so deeply entangled with the gradual development of geometric interpretation of our observations in terms of what we now call Euclidean space (and further to manifolds, etc.) that observation of noncommutative space is (for the moment) impossible for us because we have never learned to do that? The wide-ranging philosophical ramifications of this are obvious at first glance, but we do not go into this in a mathematical work. I just want to mention some aspects I learned about during recent contacts with colleagues from the mathematical physics area looking at the noncommutatve model with an open mind.

1. *A relation with string theory?* The description of an observed minimal object, a point say, in the noncommutative world, leads to a string (it may be open or closed; the isomorphisms $Y \cong SL(\Lambda_t)$ play a role here) in the space Y, for

now identifiable to usual space-time. In the moment space X we see it as a higher-dimensional string (again open or closed) and the gain in dimension relates to the use of irreversible time during observation. This situation is very similar to the basic ideas in string theory; of course we have no energy introduced in our theory yet but the stringlike properties follow from seemingly harmless axioms expressing weak continuity properties phrased in terms of posets and basic operations. An object (particle) moving between points in the noncommutative world may be represented by a string movement and carefully defined periodic movement by a vibrating string. This is just a formal fact here; whether it has a meaning for the suitability of the noncommutative model in string theoretic considerations could be an interesting question.

2. *Chaos from order.* The local injectivity, which is the unambiguity interval in T for some $\lambda_t \in \Lambda, t \in T$, is obtained as an existence statement. We did not impose a notion of size, dimension, or length on intervals in T, but the unambiguity interval may be thought of as arbitrarily small. The closed interval I_t on the other hand is big enough so that it allows us to discover all points representing the temporal points of Λ_t, that is, enough time from T to allow the construction of $\text{Spec}(\Lambda_t, I_t)$ and its spectral topology. The local injectivity describes a very local order at the infinitesimal scale, but at the scale of I_t chaotic aspects of the (topology) space derive from the fact that T-intervals contained in I_t are much larger than unambiguity intervals for elements one is considering in certain mathematical statements. These chaotic aspects thus appear when observations are carried out for intervals at the scale of the I_t, but infinitesimal observations (if one can imagine such in a real world) carry much more order; that is, they are more predictable.

3. *Noncommutativity makes you free.* This is a very amusing idea related to research of the brain functions. Of course the activities of a brain are also part of the events defining the noncommutative world now; physical aspects of neuronetworks and so forth are embedded in the spaced-time model. Some time ago some brain researchers excluded the existence of free will, roughly stated because at the moment of making a decision the past history of the activities determine the outcome. Now the philosophical remark about the development of our awareness in combination with commutative geometrical interpretations of observations would mean that one is not aware of the brain events in the noncommutative world outside the commutative shadow. This leaves room for extending or perhaps defining unconsciousness outside the commutative shadow, then looking at the string description given earlier; an event in the noncommutative world at time $t \in T$ appears as a string beginning at a later time t' in the commutative shadow. So, if decision making involves unconsciousness (as is well possible, perhaps plausible), then the interval $[t, t']$ in T and its relation to unambiguity intervals of some past "events" just may buy you the extra time to use free will without being aware of it.

Bibliography

[1] Aly, F. S., and F. Van Oystaeyen. "Hopf Filtrations and Larsson-Type Orders in Hopf Algebras." *J. of Algebra* 267(2003): 756–772.

[2] Artin, M. "Grothendieck Topologies." Cambridge, MA: Harvard Univ. Lecture Notes, 1962.

[3] Artin, M., A. Grothendieck, and J.-L. Verdier. "Théorie des topos et cohomologie étale des schémas." *Lect. Notes in Math.* 269, 270, 305, Berlin: Springer Verlag, 1972–73.

[4] Artin, M., and J. Zhang, "Noncommutative Projective Schemes," *Adv. in Math.* 109 (1994): 228–287.

[5] Asensio, M.-J., M. Van den Bergh, and F. Van Oystaeyen. "A New Algebraic Approach to Microlocalization of Filtered Rings." *Trans. Amer. Math. Soc.* 316, no. 2(1989): 537–555.

[6] G. Birkhoff. *Lattice Theory*, AMS Colloquium Publ. XXV, 1940, reprint edition 1967.

[7] T. S. Blyth. *Lattices and Ordered Algebraic Structures*. London: Springer Verlag, 2005.

[8] A. Connes. "C^*-Algèbres et Géometrie Différentielle." *C. R. Acad. Sc. Paris* 290 (1980): 599–604.

[9] Connes, A., and Michel Dubois-Violette. "Noncommutative Finite-dimensional Manifolds. I. Spherical Manifolds and Related Examples." *Notes IHES*/M/01/32, juillet 2001.

[10] Gabriel, P. "Des catégories abéliennes." *Bull. Soc. Math. France* 90 (1962): 323–448.

[11] Gabriel, P., and R. Rentschler. "Sur la dimension des anneaux et ensembles ordonnées." *C. R. Acad. Sci. Paris* 265(1967): 712–715.

[12] Godement, R. *Théorie des faisceaux*. Paris: Hermann, 1958.

[13] Golan, J. *Localization of Noncommutative Rings*. New York: Marcel Dekker, 1975.

[14] Golan, J., J. Raynand, and F. Van Oystaeyen. "Sheaves over the Spectra of Certain Noncommutative Rings." *Comm. Algebra*, 4(1976): 491–502.

[15] Goldblatt, R. *Topoi. The Categorical Analysis of Logic*, 2nd ed. Studies in Logic and the Foundations of Mathematics 98. Amsterdam: North-Holland Publishing Company, 1984, xvi.

[16] Goldman, O. "Rings and Modules of Quotients." *J. of Algebra* 13(1969): 10–47.

[17] Grothendieck, A. "Produits tensoriels topologiques et espaces nucléaires." *Memoirs Am. Math. Soc.* 16, 1955.

[18] Joyal, A., and R. Street. "Braided Monoidal Categories." *Adv. in Mathematics* 102(1993): 20–78.

[19] Kassel, C. *Quantum Groups*. Berlin: Springer Verlag, 1995.

[20] Lambek, J. *Torsion Theories, Additive Semantics and Rings of Quotients*, LNM 197. Berlin: Springer Verlag, 1971.

[21] Lambek, J. "Noncommutative Localization." *Bull. Am. Math. Soc.* 79(1973): 857–872.

[22] Lambek, J., and B. Rattray. "Localization and Duality in Additive Categories." *Houston J. of Math.* 1(1975): 87–106.

[23] Lambek, J. "Localization at Epimorphisms and Quasi-injectives." *J. of Algebra* 38(1976): 163–181.

[24] Le Bruyn, L., and F. Van Oystaeyen. "Quantum Sections and Gauge Algebras." *Publ. Math.* 36 (2A)(1992): 693–714.

[25] Le Bruyn, L., M. Van den Bergh, and F. Van Oystaeyen. *Graded Orders*, Boston: Birkhauser Verlag, 1988.

[26] Li, H., and F. Van Oystaeyen. "Zariskian Filtrations." *Comm. Alg.* 17(1989): 2945–2970.

[27] Li, H., and F. Van Oystaeyen. "Zariskian Filtrations." Kluwer Monographs, Vol. 2. Boston: Kluwer Academic Press, 1995.

[28] McLane, S. *Categories for the Working Mathematician*. New York: Springer Verlag, 1974.

[29] Manin, Yu. *Quantum Groups and Noncommutative Geometry*. Montreal: Publ. Centre Math., 1988.

[30] Murdoch, D., and F. Van Oystaeyen. "Noncommutative Localization and Sheaves." *J. of Algebra* 35 (1975): 500–525.

[31] Năstăsescu, C., and F. Van Oystaeyen. *Graded and Filtered Rings and Modules*, LNM 758. Berlin: Springer Verlag, 1979.

[32] Năstăsescu, C., and F. Van Oystaeyen. *Graded Ring Theory*, Math. Library. Vol. 28. North Holland, Amsterdem, 1982.

[33] Năstăsescu, C., and F. Van Oystaeyen. *Dimensions of Ring Theory*. Boston: D. Reidel Publ. Co., 1987.

[34] Năstăsescu, C., M. Van den Bergh, and F. Van Oystaeyen. "Separable Functors Applied to Graded Ring Theory." *J. of Algebra* 123 (1989): 397–413.

[35] Nwabueze, K. K., and F. Van Oystaeyen. "Presheaves over Lattices. Application to Twisted Limits of Finite Groups." *Comm. Alg.* 25(3) (1997): 887–903.

[36] Popescu, N. *Abelian Categories with Applications to Rings and Modules*. London: Academic Press, 1973.

[37] Procesi, C. *Rings with Polynomial Identities*. New York: Marcel Dekker, 1977.

[38] Radwan, A., and F. Van Oystaeyen. *Microstructure Sheaves and Quantum Sections over Projective Schemes*, Paris: Hermann, 1995.

[39] Rafael, M. D., *Separable Functors Revisited*, Comm. Algebra, 18, 1990, Vol. 5, 1445–1459.

[40] Sallam, R., and F. Van Oystaeyen. "A Microstructure Sheaf and Quantum Section over a Projective Scheme." *J. of Algebra* 158, no. 1 (1993): 201–225.

[41] Serre, J.-P. "Faisceaux Algébriques Cohérents." *Ann. Math.* 61 (1955): 197–278.

[42] Stenström, B. *Rings of Quotients*. Berlin: Springer Verlag, 1975.

[43] Tennison, B. R. *Sheaf Theory*, LMS Lect. Notes Vol. 20. Cambridge: Cambridge Univ. Press, 1975.

[44] Van Oystaeyen, F. *Prime Spectra in Noncommutative Algebra*, LNM 444, Berlin: Springer Verlag, 1975.

[45] Van Oystaeyen, F. "Compatibility of Kernel Functors and Localization Functors." *Bull. Soc. Math. Belg.* XVIII (1976): 131–137.

[46] Van Oystaeyen, F. *Algebraic Geometry for Associative Algebras*. Math. Monographs Vol. 232, New York: M. Dekker, 2000.

[47] Van Oystaeyen, F., and A. Verschoren. *Noncommutative Algebraic Geometry*, LNM 887. Berlin: Springer Verlag, 1981.

[48] Van Oystaeyen, F., and A. Verschoren. *Reflectors and Localization, Applications to Sheaf Theory*, Lect. Notes Vol. 41. New York: M. Dekker, 1978.

[49] Van Oystaeyen, F., and L. Willaert. "Grothendieck Topology, Coherent Sheaves and Serre's Theorem for Schematic Algebras." *J. Pure Applied Algebra* 104 (1995): 109–122.

[50] Van Oystaeyen, F., and L. Willaert. "Cohomology of Schematic Algebras." *J. of Algebra* 183, 2 (1996): 359–364.

[51] Van Oystaeyen, F., and L. Willaert, "Examples and Quantum Sections of Schematic Algebras." *J. Pure Applied Algebra* 120 (1997): 195–211.

[52] Yetter, D. N. "Quantum Groups and Representations of Monoidal Categories." *Math. Proc. Cambridge Phil. Soc.* 108 (1990): 261–290.

[53] Zariski, O. *Algebraic Sheaf Theory*. Sunrise, FL: AMS, 1956, 117–141.

Index

W

X

Y

Z

For Product Safety Concerns and Information please contact our EU
representative GPSR@taylorandfrancis.com
Taylor & Francis Verlag GmbH, Kaufingerstraße 24, 80331 München, Germany

www.ingramcontent.com/pod-product-compliance
Lightning Source LLC
LaVergne TN
LVHW021036011026
849160LV00009BA/669
* 9 7 8 1 4 2 0 0 6 0 5 6 0 *